Measurement and Meaning

Measurement and Meaning

Ferenc Csatári

LEXINGTON BOOKS
Lanham • Boulder • New York • London

Published by Lexington Books
An imprint of The Rowman & Littlefield Publishing Group, Inc.
4501 Forbes Boulevard, Suite 200, Lanham, Maryland 20706
www.rowman.com

6 Tinworth Street, London SE11 5AL, United Kingdom

British Library Cataloguing in Publication Information Available

Library of Congress Cataloging-in-Publication Data Available

ISBN 978-1-4985-8299-5 (cloth)
ISBN 978-1-4985-8300-8 (electronic)

To Ábel and Máté

Contents

Acknowledgments

I am indebted to Balázs Gyenis, András Jánossy, András Máté, Gábor Szabó, Lászlcó E. Szabó, and my anonymous reviewer for their valuable comments on the text at different stages of its development. I am grateful to Jana Hodges-Kluck and Trevor F. Crowell, my editors at Rowman & Littlefield. My special thanks to Ildikó Lakatos for her support.

Introduction

A usual way to begin a scientific book is to present some quotations, and then excuses for choosing the given topic. I would not dare to break with this nice tradition. However, I would give my obligatory citations in-text in this preliminary chapter. While I will provide a brief and rough introduction to the concept of measurement relying on the thoughts of renowned scientists, I will also explain myself by revealing my motives. I believe this will provide the needed preliminaries for outlining my argumentation, the logical structure of the book, and advancing my results.

Texts on measurement tend to begin with something like this: "measurement is of crucial importance in sciences." Indeed, it would be hard to deny, let alone that scientists usually highly esteem measurement themselves. William Thomson, also known as Lord Kelvin, put once: "if you cannot measure, your knowledge is meager and unsatisfactory" (Kuhn 1961, 161). Galileo Galilei allegedly advised that "measure what is measurable, and make measurable what is not so." This brings us to really important questions for which answers are sought in this book: what is measurable, what is it to make something measurable, and what is it to be able to measure at all?

What this book will not provide is a strict definition, it will not announce a verdict on what measurement is. It is still worth noting that while many theorists and practitioners regard the issues of measurement by and large trivial or well understood, others cannot reach an agreement on the very concept of it. What few would question is that measurement is a kind of "interface" or "mediator" between theories and direct sensual experience (Margenau 1959). But beyond this point measurement concepts diverge. Some define it as the estimation of a ratio of a magnitude of a given quantity and an arbitrary unit of the same quantity (Maxwell 1890, Michell 2007). Others maintain that when we measure we assign numbers to objects according to rules (Campbell 1920, Stevens 1946, Narens 2007). Some claim that measurement is intersubjective and objective evaluation (Mari 2003). Some define measurement "as a comparison to a generally accepted standard" (*Scientific American* 2004). Some even venture that "measurement is an operation that locates an item in a logical space" (van Fraassen 1980, 164). Still others insist that we do not need a stiff definition. One recognizes measurement once she got acquainted with the large family of representations which qualify as such (Krantz et al. 1971).

Though below I will provide a detailed conceptual analysis of measurement, as said, I do not devote myself to a dedicated definition either. Nonetheless, it is worth making one thing clear at the beginning: even if I place measurement on the front line of empirical experience, along with Thomas Kuhn I would refrain from calling "any unambiguous scientific experiment or observation" a measurement (Kuhn 1961, 162). We gain nothing by so dissolving the notion. To say the least, measurement involves relations, ratios, numbers, or structures. Thus, to my view, to detect foaming in a test tube after mixing hydrogen peroxide with yeast or to observe frog legs to jerk in a circuit is not what we should essentially call measurement. But let it: trivially, we can order the number one for a jerking leg and zero for a non-jerking one, and now we have arrived at something like a scale.[1]

Putting aside these moderately fruitful terminological speculations, let us observe what is at stake. Measurement can be perceived as a procedure for exhibiting structures in empirical data. It is an old philosophical issue how structures are given for us in our experience. Roughly, some say they are "there," while others insist they are projected onto it by "ourselves." Thus by exhibition someone can mean revealing some given feature, but she likewise can refer to a creative procedure. I think the problem is quite complex and I do not give my bets at the start. Indeed, measurement can be regarded as the instrumental approach to epistemology, where what is given and what is achieved both have major words. The concept likewise encompasses the theoretical grounding of procedures for quantification, the actual experimental aptitude for showing up quantities, the institutional-instrumental endeavors for calibration, and the individual measuring acts of engineering or social data collections.

Before giving the reader more hints on what she may expect in the following pages, a few words on my motives to write this book are in order. Having been trained in logic and philosophy of science, I have long been occupied by the general problems of epistemology and semantics. In particular, I have long been wondering what is it to hold a realist (or antirealist) stance about empirical statements. This query traces out the main axis of the study. Every historical or conceptual episode is meant to further the reader toward my semantic conclusions. But there is another side of my interest, stemming from my past (and rather limited) experiences as a practicing researcher in sociology. Muddled methodology and incoherent implications that featured some—by no means all—research projects I met along the way often came to me as stunning.

This book, of course, is not meant as a catalog of measurement methods. My approach is rather a generalist one in that I will focus on some common issues of primal philosophical interest, while trying not to drift far from an everyday understanding of measurement. Thus I will mention particle physics measurement, quantum physical measurement, factor or big data analysis only in passing if at all. Rather, I will provide an introduction to the so-called measurement theories, that is, the foundation projects for measurement and a hopefully deep enough conceptual analysis thereof. Still, whenever needed I will underpin my argumentation with minor case studies of practical procedures.

My aim is to end this book with a plausible semantic interpretation for a special class of assertions with empirical content: statements of measurement results. On the way, some of my observations and results are negative. I will show that the classical, empiricist grounding of measurement is untenable since it is burdened by circularities. I will also argue that the practice of measurement in social sciences is often opportunistic and lacks cogent conceptual grounding. Further, I will insist that holding a realist view on quantities is awkward. In particular, given a quantitative property p, a conventional unit and a real number x, holding that "p is x units" is true or false independently of our knowing which is likewise uncomfortable and unnecessary.

Nevertheless, there will be a positive side for each of these points. I will argue that—along with possible theoretical, instrumental, and institutional machinery—the operational exhibition of congruent phenomena may provide the needed foundations for measurement. Despite idiosyncrasies, searching for these patterns is also open in social sciences. All in all I will argue for a constructivist approach to measurement,[2] and suggest that researchers can take care of meaningful and cogent measurement operations.

On the other hand, I will insist that measurement procedures cannot exhaust the meaning of concepts of quantities. Meaning cannot be equated with truth conditions either. It may well happen that we cannot state the exact conditions for entirely meaningful statements. After looking into some accounts of errors and uncertainties, I will show that both consistent realism and plausibly interpreted operationalism gets in trouble when it comes to truth: statements of measurement *always* comes out as false for the first and true for the latter. A way out from this muddle is to take a constructivist approach. Instead of the realist credo, and drawing somewhat on operationalism, I will hold that given a quantitative property p, a conventional unit and a real number x, we may assert that "*p* is *x* units" is true or false in virtue of our valid measurement procedures. Note that this means that the principle of bivalence, that is, that every well-formed statement is true or false independently of our knowing which, cannot be held anymore.[3]

The text is structured as follows. Chapter 1 gives an overview of the early representationalist theories of measurement. Far beyond mere historical interest, this part presents an introduction to the main measurement theoretical problems, and also to their conceptual background. It also introduces a closely related still rival (at times insurgent) approach: operationalism. It will be revealed that operationalism cannot be regarded as a monolithic view as it exists in different flavors, and it is even troublesome to unfold it as a systematic standpoint.

Chapter 2 pursues the issues introduced earlier into deeper philosophical analysis. I will take up the problems of rules and concatenation and investigate the role of conventions and phenomenal congruence in quantification. I will also dive into the muddy waters of realism. After investigating what is to hold a realist stance, I will argue for a constructive approach instead. Most importantly, I will promote the intuitionist mathematics (and logic) as more apt for describing empirical data than the classical one.

Chapters 3 and 4 delve into the issues of physicalism and measurement in social sciences. Also, they introduce the axiomatic approach and the problems of invariance and meaningfulness. The axiomatic theory, a monumental foundationalist project, develops a clear and sophisticated notion of representation, and it is also keen on structures featuring the different measurement operations. In spite of its significant intellectual achievements, axiomatic representational theories never made the historic breakthrough that they should have—it is widely held. Some of the reasons are observed in chapter 4 along with some typical conceptual problems of operationalization. I will also provide illustrative minor case studies. These sections add quite a lot to the whole picture but the main argument of the book can be followed without them. So readers uninterested in social science issues and axiomatic methods may even skip them. But only on their own responsibility. I would not advise to deprive themselves of a pleasant experience.

After a detailed analysis of an error theory of a realist nature, and a critical survey of what I call constructionism, chapter 5 draws the conclusions of these investigations. In particular, I will argue that we must take a constructivist approach when it comes to the truth of measurement result statements. These statements are not true or false by themselves, but only in virtue of our valid procedures. Apart from certain institutional conditions, establishing validity always contains operational elements. Exhibiting congruent phenomena often lies at the heart of validity.

NOTES

1. It is not at all to say that I would by default dismiss other than numerical structures when talking about measurement. Not the least, as we will see, it is entirely unclear how to interpret numerical at all.

2. Since constructivism or constructionism has a glittering career as a keyword in the recent decades, there is a clear need for setting my viewpoint apart from some popular trends, and I will do so accordingly toward the end of this book. Still, it is worth noting in the beginning that my constructivism has little to do with the sociology of scientific knowledge. Rather, it has its roots in intuitionism and constructive mathematics and in their subsequently induced movements in semantics and metaphysics.

3. What validity lies in is, of course, an important question, and I will not claim to have a full-fledged answer. I will indicate, however, that further analysis of phenomenal congruence may prove to be a fruitful direction.

Chapter 1

Empiricism

It is widely agreed upon that quantitative concepts are essential to sciences. Other types of concepts, though also useful (or even indispensable), have fewer merits. Qualitative concepts, for one, assign properties to objects, or more precisely, assign objects to classes, but remain silent about everything else; in other words, they cannot tell anything about the degree of part-taking. Debbie is a witch, Debbie belongs to the class of witches. Comparative concepts get farther: they can grab some relations between certain objects with regard to certain properties. Debbie is more wicked than me, Debbie's nose is harder (and more magnetizable) than mine. But they do not provide room for the degree of differences.

Quantitative concepts allow for the description of the properties or relations of objects in numerical terms. One can measure the *length* of Debbie's nose, the length of my nose and she can even compare the two values and determine their difference. Quantitative concepts have a "higher status" than those of classifying or relating concepts in that they make a sharper description of the world possible. In fact, they are the ones of real interest for scientific inquiry, and a major goal of every scientific theory is to define the qualitative concepts sufficient for empirical confirmation.[1]

The appropriate empirical observations are realized, to a large extent, via measurement. The purpose of measurement is to provide the quantitative concepts, that is, certain *magnitudes*, with numerical values. Generally speaking and according to a wide consensus, *measurement assigns numbers to properties of objects or subjects of study*.[2] In addition, it is also widely held that this assignment should not be arbitrary, ordering numbers to quantitative concepts requires certain rules—and may require different set of rules depending on the given measurement procedures.

From a reverse point of view, one can say that by measurement objective properties, or (more cautiously put) empirical data are numerically represented. This is why theories stemming from this sentiment are called representational measurement theories. I will refer to an early breed of these theories as empiricist. These theories, which lay stress on the empirical rules of measurement, are the subject of this chapter. By contrast, axiomatic representational theories focus on structures, and I will deal with them later in this book.

Again, also in this chapter, I introduce the extremely empiricist standpoint of operationalism. Untenable though as a theory of meaning, its contributions to the modern accounts of measurement, and science in general, are of utmost importance.

1.1 COUNTING

Though many of us may refrain from calling counting measurement in an everyday sense, we have good reasons to do so still. Counting usually fits in well with the theoretical accounts of measurement—and practice, too. Also, more complex measurement operations usually rely heavily on counting. We may even venture that counting lies at the heart of every kind of measurement.

Gottlob Frege was the one to first realize that a "statement of number" is an assertion about a concept (Frege 1884).[3] Numbers are not abstractions from things nor properties of them like shape or color. Instead, when I say, "There are five witches flying around in the hall on their brooms," I assign the number five to the concept "witch flying around in the hall on her broom." Frege arrived at this thought by observing that the same phenomenon can be numerized in different ways depending on its conceptualization. I can talk of five witches, a pack of witches, five brooms, two crooked brooms and three straight ones, and so on in the very same perceived situation.

In other words, given a *C* concept, we can ask: "How many *C*s are there?" We can answer such a question by applying numbers in our response. Numbers so conceived are called cardinal numbers, as they determine the cardinality of the class of objects falling under *C*.

When are we justified in saying that the numbers for two distinct concepts are equal? The answer is that they are equal whenever there is a one-to-one correspondence between them. That is, *C* and *D* are *equinumerous* (have the same cardinality) iff[4] there is an **R** relation so that it relates every *c* in *C* to one and only *d* in *D*, and for every *d* in *D* there is one and only one *c* in *C* which is **R**-related to it. Actually, Frege used the principle that the number of *C* and the number of *D* is equal iff *C* and *D* are equinumerous in the very definition of number, and called it Hume's principle.[5]

From a measurement point of view, counting is the assignment of integers to (the concepts of) objects, or, to be more precise, the assignment of non-negative integers. It may not be trivial for everyone that we can count zero of something, but this assumption is rather fruitful, and anyway, it is indeed a fact that we are able and willing to express thoughts like "there are no witches in the hall." There is of course a relation which uniquely relates each element of the empty class to the elements of class of witches in the hall (if there is none): the empty relation. (We can also put it this way: given a class A so that a is in A iff a is a witch and a is in the hall, A is empty.)

Carnap argues that event series are at the very foundation of counting procedure, namely, the series of pointing events (even if adult human counters do not need pointing in most of the cases anymore) (Carnap 1966). When we count, we establish one-to-one correspondence between the pointing events and the counted objects, that is, a bijection between a class of events and a class of objects.

Other series of events different from pointing acts are also suitable, of course. Finger counting, sometimes regarded as the very base of human number concept (Csatári 2008), establishes correspondence between objects and displayed fingers. Enumerating the number names, either aloud or silently, also suits as a class of events to correspond to another given class of objects or events. Automatized counting mechanisms in computer systems, aptly called counters, also relate one event class (the increment of the counter) to another (the realized instance of the observed event type, e.g., the transfer of a specific package).

But at this point we may faintly smell a bit of circularity. If we measure by events (e.g., pointing gestures), how are we to assess the numerical differences between the groups of pointing gestures? By further rounds of pointing gestures? At the end of the day, the problem comes down to this: how do we have access to numbers, how are they given to us. No easy answer can be given here. Indeed, Frege's whole project was after showing that numbers (and arithmetic) can be derived from basic logical principles.[6] Others insist that relying on some kind of intuition is inevitable. It is definitely not the place to settle this issue.

What is sure, simply telling *what numbers are* does not directly bring us to the solution of this philosophical problem. Still, getting on with a clear definition, say, according to which (natural) numbers are sets containing each lesser number, has benefits for us. It clearly reveals that by counting we measure by sets.

A final note. For reasons yet to become clear, it is important to stress again: with counting we do not assign numbers (sets) to objects, but we assign them to sets of objects (where a set may even happen to contain one or none of the objects, of course). When counting a soccer team we do not give number one to the goalkeeper, number two to the first back and so on up to the eleventh team member, however, it *is* one possible one-to-one correspondence relation (setting apart the fact that eleven as a set does not contain eleven, but does zero). Instead we assign the number eleven to the whole team, that is, to the class of all teammates on field. Independently of this, we may assign numbers to the team members and write them on their jerseys, but this procedure clearly does not count as counting. We may possibly call it tagging. Though it may sound trivial, this is a crucial issue in the use of numbers as we will see when discussing the theory of measurement levels.

1.2 ORDERS, SYMBOLS, MAGNITUDES

A next step toward a more elaborated measurement concept is the establishment of orders. Our past counting acts (as series of events) do exhibit an ordering, namely, they can be ordered by the numbers of the occurrences of pointing gestures. Human memory is not the most reliable tool to compare these past acts, we need some more permanent representation of them. To record the different "rounds" of pointing acts, we can use, say, tally marks, streaks made on a paper or on a beam in the local pub. If we compare these rows of tally marks, we find that their different groups can be linearly ordered naturally by their cardinality. Formally speaking, there is a relation **R** so that if *a* and *b* are tally mark groups, the following applies:

(i) $a\mathbf{R}b$ or $b\mathbf{R}a$,
(ii) If $a\mathbf{R}b$ and $b\mathbf{R}a$ then $a = b$,
(iii) If $a\mathbf{R}b$ and $b\mathbf{R}c$ then $a\mathbf{R}c$.

In other words, the relation **R** is (i) total, (ii) antisymmetric, and (iii) transitive.

Tally mark groups can be regarded as genuine *numerals*. In turn, numerals are nothing else that the physical representation of numbers (and let us stick to the integers for the time being), by convention. In turn, numbers, as we agreed, are sets constructed by specific rules. One may find that tally marks, as conventional symbols, still bear some strains likewise related to the empirical process of counting as to the formal procedure of the construction of numbers. But as we arrive to more abstract graphical representations, so will the conventional character of these symbols become obvious.

The British physicist and philosopher, Norman Robert Campbell, one of the most prominent figures of the early representational measurement theories, usually talks about numerals in the context of measurement; according to him measurement is "the assignment of numerals to represent properties" (Campbell 1920, 267).[7] The order of numerals is an "artificial" order, since the order of the graphical symbols for representing numbers is based on mere convention (the systematic rules for creating new numerals from the symbol kit are also conventional). But natural orders are abundant wherever we look, and, what is more, underlie conventional orders: "order may also arise, not from an arbitrary convention, but from real properties of the things ordered; and it is of course the existence of this real order which has led to the invention of arbitrary orders to denote the things characterised by it" (ibid., 270). For instance, the row of the houses in my street, a train's progress from *A* to *B*, the vertical change in the density of the atmosphere all exhibit natural orders. According to Campbell, the order of numbers (or as he sometimes puts: "Numbers"), conceived as classes of classes (in particular: sets of sets), is also natural, not in the least because they are to be regarded as a genuine physical quantity: "number, the physical property, is represented by numerals in just the same definite way as weight is represented" (ibid., 296). Traditionally, such orders are often denoted by numerals, however, numerals are also often given roles where no relevant order is involved: for example, in the case of football players, registration numbers, the lottery, or washing machine programs.

Now what is the order these natural phenomena exhibit? Campbell maintains that they are ordered by a transitive and asymmetrical relation (ibid., 270). For such a relation we should take our above transitivity rule (iii) together with the following:

(iv) If $a\mathbf{R}b$ then not $b\mathbf{R}a$.

Now, it is clear that the above order of numerals is of a different kind. The reason is that equality is not yet accounted for in this setup. This immediately brings us to the question: how should we, then, assign numerals to objects (properties)?

We can go the following way. We can assign different numerals to the same entity, iff the following holds for numerals a and b: $a\mathbf{R}b$ and $b\mathbf{R}a$. It is intuitively clear that we can assign numerals which are different physical representations of the same number to the same thing. To put it another way, we can define a partition on the class of numerals through its equivalence classes. The so-gained class has an ordering analogue to Campbell's natural order, plus the totality requirement, (i). A possible "ideology" behind adding totality is to restrict the class of natural orders to a given quantity. However, whether these structures will be analogue in every essential respect is an important question. It will be discussed at length later accordingly.

The above characterization of a quantity is not universally held. For instance, according to Russell the definitive difference between *quantity* and *magnitude* can be seen as follows:

> There are a certain pair of indefinable relations, called *greater* and *less*; these relations are asymmetrical and transitive, and are inconsistent the one with the other. Each is the converse of the other, in the sense that, whenever the one holds between *A* and *B*, the other holds between *B* and *A*. The terms which are capable of these relations are *magnitudes*. Every magnitude has a certain peculiar relation to some concept, expressed by saying that it is a magnitude of that concept. Two magnitudes which have this relation to the same concept are said to be of the same kind; to be of the same kind is the necessary and sufficient condition for the relations of greater and less. When a magnitude can be particularized by temporal, spatial, or spatio-temporal position, or when, being a relation, it can be particularized by taking into a consideration a pair of terms between which it holds, then the magnitude so particularized is called a *quantity*. Two magnitudes of the same kind can never be particularized by exactly the same specifications. Two quantities which result from particularizing the same magnitude are said to be *equal*. (Russell 1903, 167)

It follows that two magnitudes can never be equal; what can be equal is their particularization, quantity. That makes the discussion on measurement a bit circuitous. That is why I am rather adopting Michell's approach (Michell 2007): a quantity is a class of all magnitudes of a certain kind, that is, a magnitude is an instance of a quantity. Nevertheless, we can still regard magnitudes as unique (as some ideal entities) if we like by saying that two individual properties are of the same magnitude when we measure equality. Contrarily, we can likewise say that two magnitudes are equal. As it is just a matter of ways of speaking basically, I will only be strict in this respect where it is really necessary.

According to Campbell, natural order features all that is measurable. By systematically putting objects in the two pans of a balance scale, we can determine their natural order with respect to their weights. On the other hand, the shape of the measured objects are not a property exhibiting natural order (we cannot say seriously that a box is more of a sphere than a pyramid), thus cannot be an object for measurement. Campbell gives the example of color as an immeasurable property (Campbell 1920, 272). True, colors can be traced back to the frequency, intensity, and combination of light waves, that is, to measurable quantities. Though this strong *theoretical* armature results in (some esoteric combinations of) numbers, they do not exhibit a natural order. We may talk about things that are redder than others, but on the one hand we are not ready to assess whether a green or a blue shirt is "redder," and, on the other, we may also be perplexed to decide which of two red shirts is redder. According to Russell here we are not dealing with the relation of greater or less, but that of *resemblance* (Russell 1903, 171).

A borderline case often mentioned by Campbell is hardness. Mohs' scale of hardness exhibits an order of minerals by the possibilities of scratching one with the other. If mineral *a* scratches mineral *b* then mineral *a* is harder then mineral *b*. When finding that in the above case *b* cannot scratch *a*, and, further, having a mineral *c* so that if it scratches *a* then it also scratches *b*, then we have standard scale for the hardness of minerals.[8] Now the problem is that this empirical requirement does not hold in each case. Further, if it held, we would have an order only without any units to determine differences (or distances) in hardness—a clear expectation for a measurable quantity as described in the next section.

1.3 EXTENSIVE AND INTENSIVE MEASUREMENT

The distinction of extensive and intensive quantities is a historical one, it can be traced back to Leibniz and Kant. A large scale of quantities is traditionally described as extensive, such as length, mass, or speed. To characterize them, let us take a look at the *rules* for their measuring procedures. According to Carnap (1966, 63), the first rule for extensive magnitudes concerns with equality, a concept we missed so far (Carnap 1966, 63):[9]

(i) If $E_M(a,b)$ then $M(a) = M(b)$.

This rule suggests that whenever we have an empirical procedure *E* by which we can establish equality between two object *a* and *b* in respect of some of their properties, then the value of the respective magnitude *M* will be the same for both objects. Putting two weights in the different pans of the balance scale, if we cannot realize any bias, the two objects have the same weight.

The second rule introduces additivity. Whenever two objects are combined in a specific way, their values in a given quantity is summed up, that is, the resulting value of the given magnitude is the arithmetical sum of the values of the two original magnitudes. Given a rod *a* with a length *l*(*a*), and given another rod *b* with the length *l*(*b*), if we put them together end to end by a straight line, the length of the "new" object, *l*(*c*) will be the sum of the fist two: $l(c) = l(a) + l(b)$. Similarly, if we measure objects *d* and *e* on a spring scale resulting in weights *w*(*e*) and *w*(*d*) respectively, when putting them together on the scale we will get a value $w(f) = w(d) + w(e)$. (We also have a procedure for the a balance scale, but it requires a bit more tinkering and the third rule, see below.)

The above examples suggest that the legitimate ways of combination is highly different for the different measured magnitudes of objects. In case of weights, we require the objects of measurement to hold a specific spatial position, in particular, to be in the same pan of the same scale when combined. In the case of length, we need a straight line (a geodesic) to talk about a proper combination. Of course, it is easy to see that having two adjoining line segments (*a*, *b*) and (*b*,*c*) with an angle other than 180 degrees at their meeting point *b*, we will have a triangle (*a*,*b*,*c*) and thus the hypotenuse will be shorter than the two legs taken together.

The appropriate way of combining objects to abide the rule of additivity regarding a given quantity is called *concatenation*. In the literature, the symbol "$\circ$" is often used for denoting the empirical procedure of concatenation, and I will follow this convention here.

Thus, our second rule can be stated formally as:

(ii) $M(a \circ b) = M(a) + M(b)$.

The third rule is the unit rule. The rule states that in case of an extensive quantity it is assured that any two (finite) differences of the given magnitudes can be compared. For this, it is enough to specify the empirical conditions ED for regarding two differences of a given quantity *M* as equal:

(iii) If $ED_M(a,b,c,d)$ then $M(a) - M(b) = M(c) - M(d)$.

The rule can be aligned with by establishing a unit. Having a large set of weights, we may chose one of them as a unit. It is worth calculating with the typical "size" of the differences we are about to measure, and choose the unit so to be significantly smaller than that. Then, laying the measurable weights (*a* and *b*) in the two pans of a balance scale, we can count how many copies of the unit weight have to be placed in the upper pan (thus we use the relevant concatenation operation) to reach a balance between the pans again. Then we can take other measurable pairs (*c* and *d*) and so on. Finding, then, that we place equally as many copies of the unit weight in different cases, we can regard the empirical condition (iii) fulfilled.

The establishment of a unit is done in practice either by a standard prototype, or by reducing it to some known phenomena of stable behavior usually described by already established measurement procedures of different kinds. For instance, the unit of mass[10] is specified by a standard prototype, a platinum-iridium cylinder stored at the International Bureau of Weights and Measures in Paris. Some decades ago, the unit of length was also specified by a similar object stored at the same place: a platinum-iridium bar, whose length was regarded as 1 meter at the melting point of ice. But nowadays, the definition goes like this: "The metre is the length of the path travelled by light in vacuum during a time interval of 1/299 792 458 of a second" ("SI brochure" 2006, 112). In turn, the recent definition of a *second* is the following: "the duration of 9192631770 periods of the radiation corresponding to the transition between the two hyperfine levels of the ground state of the caesium 133 atom" (ibid., 113).[11]

The examples above suggest that defining a unit is not a trivial task. When using prototypes, the following problems arise. The *étalon* object's properties may change in different circumstances. For this reason, it is regarded as standard only in specific conditions, for example, the meter bar is regarded "calibrated" at 0°C. Nevertheless it is hard to be sure that one can take all of the relevant factors into account, even just in principle. That is why we do not even state that various electromagnetic forces or natural background radiation do not have their effect, we only regard them as negligible in normal circumstances.

Another problem is the portability of the prototype unit. Of course, it is not possible to move along the prototype object from one measurement site to another. We must make replicas. But these copies cannot be identical with the prototype in every respect, and, equally important, the standard conditions cannot be maintained anymore outside isolation.

The method of measuring stable phenomena (as the propagation of light in vacuum, or the behavior of caesium 133) seems to solve the problem of portability. Unit defining procedures can always be performed at any place, at least in principle. In reality, these are highly sophisticated measurements themselves, requiring a good deal of well-calibrated equipment—and at this point we are back at something very similar to the problem of portability of prototypes. Worse, with account for relativistic effects, further issues arise. More to be said on this in the next chapter.

What is to be seen here is that unit definitions often involve the use of already established units of different kinds, thus export the load of standardization to the realm of other quantities. Indeed, in the modern standard metric system (SI) the definitions of the seven base units are strongly interwoven. As we have seen, the definition of meter relies on that of second. Ampere and candela (the unit of of luminous intensity) are defined through time, length, and mass. Mol (the unit for the amount of substance) refers mass in its definition. In turn, mass has a unit prototype (ibid.). Eventually all but one of the units are based either on the standard prototype of mass, or on the standard unit of second. What is left is kelvin, the unit for the archetypal intensive quantity: temperature.[12]

Traditionally, those quantities are regarded as intensive, which exhibit order, but lack additivity. If "an intensive magnitude can be perceived in an object as greater than, less than, or equal to the magnitude of the same property in another object, yet we cannot assign a ratio to two unequal magnitudes" (Castellano 2007), then we can say that extensive magnitudes are those for which we can assign that ratio. In practice, though, we are prone to presuppose much more of intensive quantities than mere order. In the spirit of what is said above we could better say that intensive quantities are those, on which no (meaningful or natural) concatenation operation can be applied. While, as we have seen above, length and time can be regarded as extensive quantities, intensive ones are exemplified by temperature or beauty—though the latter is not regarded as measurable in the society of morose scientists nowadays. A further feature often attached to these quantities is that, typically, they are measured indirectly. But as we will see, by this property we cannot really set them apart from extensive magnitudes.

The measurement rules for intensive magnitudes are somewhat different from those of the extensive ones. Let us turn now to these rules through the example of temperature (Carnap 1966, 62–69). Our first rule for measuring temperature is similar to the first rule of extensive measurement we have seen above:

(i) If $E_T(a,b)$ then $T(a)=T(b)$.

That is, we must have an empirical operation for establishing equality between the temperatures of two given bodies. In the most simple case we can touch them, but it can be rather painful sometimes. It may be more comfortable to establish equality between their infrared radiance visually.

But now we must introduce a new rule for the one-way difference. We did not need this step in case of the extensive magnitudes, since concatenation took care of it by guaranteeing the growth when putting two magnitudes properly together. Our new rule, introducing the concept of "larger than," goes like this:

(ii) If $L_T(a,b)$ then $T(a)>T(b)$.

It is a modest requirement that when we cannot establish equality by the above empirical operation, we should be able to establish not mere difference, but also tell which of the bodies is hotter.

Let us see now the remaining three rules:

(iii) If $D_{T_0}(a_0)$ then $T(a_0)=0$.
(iv) If $D_{T_1}(a_1)$ then $T(a_1)=1$.
(v) If $ED_T(a,b,c,d)$ then $T(a)-T(b)=T(c)-T(d)$.

The last one, rule (v) may be familiar, and also a surprise to see it here. It postulates an empirical operation for establishing equal differences for the given magnitude. But in order to be able to talk about these differences in numerical terms, we must have a unit concept. Setting up one for temperature seems to be a bit trickier than for the extensive magnitudes even at a first glance: we cannot simply take our favorite rod or pendulum. Instead, we are obliged to pick some stable, temperature dependent phenomena in the world, and characterize them with dedicated magnitudes. Traditionally, in the

Celsius scale we order to the freezing point of water (or the melting point of ice) the zero value, in line with rule (iii). Then, on the same scale, we assign another number to the boiling point of water (on the sea level), according to rule (iv). Thus, we are in possession of an interval, a unit, which we can divide into equal parts just as we like—in our case we divide by 100. Technically, we can lead these procedures the following way. Take any material that changes its volume depending on temperature. Of course, it is worth choosing one with relatively large volume differences for the sake of the ease of detection, we can also use some tricks to enlarge them, for example, we often put some mercury in a narrow glass tube. Now just mark the level of mercury first at the freezing point then at the boiling point of water, then divide the so gained length into 100 equal parts, and we are immediately at something like the Celsius scale.

It seems more or less simple. It is to be noted however, that establishing fix points and keep them fixed is not a trivial task at all. The "boiling point" of water may differ by pressure, matter of container, solute impurities. And "boiling" itself is also a vague concept.[13] Apart from this, two other serious problems arise here. First, we may ask, are we really free in choosing our material with which we establish temperature scales? Do different materials behave the same way in that we can produce similar scales with them—where similar means that they can be linearly transformed into each other, just like the Celsius and Fahrenheit scales? Second, clearly not independently of the first problem, how can we legitimately extend our scales beyond the interval defined by the two dedicated temperature points? All these problems will be addressed in the upcoming sections.

In addition to setting apart extensive and intensive magnitudes, let us observe another traditional distinction, made first by Campbell, namely that of fundamental and derived quantities (Campbell 1920). Magnitudes of fundamental (or basic) quantities, like length, mass, or time can be accessed directly by a so-called *standard series* where a standard unit is taken *n* times.[14] Putting this series against some fundamentally measurable magnitude in the appropriate way its measure can be estimated. At the same time the so-called derived quantities, such as constant acceleration or density, can be calculated only if they can be based on the measurement of basic magnitudes by discovering the concerning *scientific laws*. Thus, acceleration is obtained as the change of velocity in a given time interval, while density as the quotient of weight (mass) and volume (i.e., eventually length). Campbell also calls the two types A- and B-magnitudes respectively, and sometime refers the latter as *qualities* (Campbell 1928)—I will not follow him into this terminological muddle on these pages.

Campbell's main point is that derived magnitudes do not possess the required numerical properties when measured directly. For instance, density can be measured by observing how different solid bodies behave in different liquids. If we have a body that sinks in liquid *a* but floats in liquid *b*, we conclude that *b* is denser than *a*. Having another body that sinks in *b* but floats in *c* we can conclude that *c* is denser that *b* and it also denser that *a*. Thus the observed liquids are ordered according to their density, and we can even assign numbers to them preserving this order. But it must be seen that with this, hardly more is achieved than in the case of Mohs' hardness scale. Of course, density (of liquids) are often measured with floats with scales on which the submerge of the float in different liquids can be measured. It can be argued that by this method a measurement similar to that of temperature is accomplished (see below).

But after all, density is traced back to genuine, additive extensive magnitudes. Thus, eventually, we can order numbers to different densities which are not entirely arbitrary but bear the properties of the underlying extensive scales. In these terms we can meaningfully assert that mercury is thirteen times denser than water. Campbell insists that this reduction is of purely empirical nature; scientists had to verify the fact by experimental methods that when we double the weight of some material, volume will also be doubled.

In contrast to this, I am inclined to think that this reduction is of purely definitive nature. Not in the least, the above "law" is not even true. Taking any gas on earth, when we double their weight, we may realize a volume growth lesser than double. At the same time we may declare a growth in the density of the observed gas. And for this we can hold responsible only the very fact that density is defined by the quotient of weight and volume by scientific consent. Yes, we can save the results of our experiments by applying different theories in our explanation. Some considerations on gravity will explain our unexpected volume growth. But this is more than raw observation.[15]

In addition, as Ellis holds, many quantities are "derived" in the sense that they are measured indirectly, through the measurement of another quantity (Ellis 1966). Thus temperature is often measured via measuring the level of mercury in a tube, in other words, measuring length; velocity is often measured by gauges based on potential differences (in cars) or by Doppler-effect, in other words, wavelength shift (in case of cosmic objects). Practically speaking, quantities are rarely if ever measured directly and probably every quantity can be measured through measuring one or more other quantities. As we have seen, many of the basic quantities in modern physics are not even basic in Campbell's sense because they are *defined* by others. This leaves us with only three fundamental quantities where direct measurement could play the peremptory role. And even less after the proposed SI redefinition in 2019.

At the end of the day, it is entirely not clear how "direct" is to be understood. Could reading a gauge be counted as direct measurement? How about some optical laws and theories of perception? Or let us take the example of concatenation by a standard rod (if length, as defined by time, is ever to measured directly). Cannot determining a straight line or measuring angles at the joint points of the standard unit instances be regarded as measurement themselves? All in all, it seems to me that the distinction of fundamental and derived kinds of measurement is not so sharp as it may seem at first. Indeed, it is a perplexing one. Rather than helping conceptual clarity, these categories bring some fundamental problems to surface.

1.4 OPERATIONALISM

We cannot go on without introducing "an important footnote" (Bergmann 1956, 41), a view or bunch of views, which was, at least initially, regarded by the logical positivists as the practitioner scientist's version of their doctrine (Hempel 1956): operationalism. All the more so, as this sentiment likewise plays an immense role in later proceedings and the original findings of this study.

This "ism" has its origins in Percy Williams Bridgman's book, *The Logic of Modern Physics* (Bridgman 1927). Bridgman, a Nobel-winning physicist studying high-pressure processes, maintained that the only way for us to know the meaning of a concept is to have a way for measuring it. As Chang notes, this view is deeply rooted in his own experiences as an experimental physicist: by reaching higher and higher pressures the so far applied measuring equipment crashed, thus newer and newer ones had to be made up (Chang 2009). He also payed particular attention to Einstein's special relativity theory with its explicitly established method for ascertaining simultaneity between events far away in space by sending light beams. Thus we arrive at a new concept of simultaneity essentially different from our usual ways of ascertaining simultaneity between events happening in the same place (i.e., "very close").

Bridgman draws the example of length to show that we must face serious difficulties concerning even such prosaic concepts. His point is that we can meaningfully talk about different orders of length only with respect to their measurement procedures. Thus, measuring a wall with a measuring tape, microscopic objects by an eyepiece reticle, interstellar distances in light-years by some sophisticated speculations on spectrum shift all constitute different length concepts. Bridgman allows the possibility to call all of these quantitative concepts by a common name, length that is, if the different procedures are congruent on the overlapping ranges, that is, they provide similar results. But it is only a shorthand; we must always be aware of the different concepts based on the different measurement methods, otherwise we may easily run into pitfalls. For instance, extending the concept of length to the atomic or subatomic range without having an appropriate procedure of measurement would be simply meaningless. It is not clear what we should mean by the "inside" of an electron, or by the distance of two particles; after all it is not clear what do we mean by length in this realm.

What Bridgman highly appreciated in the spirit of special relativity was the insight that we must reinvent our basic concepts in the light of the respective empirical operations. Likewise, he was broadly content with the development of quantum mechanics, where reflections on measurement procedures play a central role in the theory. On the other hand, remaining sternly faithful to his own convictions, he criticized general relativity for its "uncritical, pre-Einsteinian point of view" (ibid.), dealing with concepts without immediate operational interpretation.

It is indeed spectacular that Bridgman's operationalism has a spirit akin to logical positivism. According to the latter, aside from the analytic statements of logic and mathematics conceived as mere tautologies, only those statements can be regarded as meaningful, which are available for empirical confirmation—at least, in theory. For the empiricist, meaning is constituted by the state of affairs where the given statement is true. For the operationalist, meaning is determined by the actual procedures of measurement. Moreover, it seems Bridgman also walks hand in hand with logical positivism when it comes to the demarcation problem. He writes: "many of the questions asked about social and philosophical subjects will be found to be meaningless when examined from the point of operations. It would doubtless conduce greatly to clarity of thought if the operational mode of thinking were adopted in all fields of inquiry as well as in the physical" (Bridgman 1927, 30–32). In a sense, this approach is stricter than that of logical positivism. For the latter *in principle* verifiability is a sufficient criterion for meaningfulness, but Bridgman presses for concrete procedures.[16]

Now it is clear that this narrow and strict version of operationalism,[17] as it became increasingly clear for the logical positivists themselves, is hardly tenable for several reasons. First, many authors in the philosophy of science teach us that scientific theories are full of unoperationalizable (theoretical) concepts, and our observations themselves are theory-laden.[18] That is, not every useful concept is operationalizable and theories cannot be avoided when measuring. In fact, Bridgman did not deny the use of theoretical concepts, he required only from a theoretical system to touch empirical operations somewhere. However, his insistence on the operational foundations of concepts and skepticism about closing up operationally different concepts were trivially at odds with a need for conceptual unity. Hempel insisted that an operational proliferation of concepts is not only practically problematic but leads to the fragmentation of science (Hempel 1966, 91–97). According to him, concepts constitutes the knots in the network of scientific knowledge linked together by the threads of theories. Scientific progress often requires the reconsideration of theories, which goes hand in hand with that of concepts. The inflexibility and plurality of operationally defined concepts thus are against the very nature of science (Chang 2009).

Second, not unrelated to the problem above, it also seems to be rightful to insist that operational procedures cannot exhaust the meaning of a concept. To begin with, they fail to give meaning for theoretical concepts and substances: only the properties of electron have meaning, not the concept of electron itself.[19] Not in the least, were there no more to the meaning than measurement operations, it would be non-sense to talk about the validity of these procedures. We would have to take them as they are, and regard them as mere tautologies (Gillies 1972, 6–7). What is more, we must face an awkward issue with semantic values also: what is it for a statement on measurement results to be false? If a quantity is no more than the way we measure it, how can we regard a measurement result faulty?[20] This suggests that no definition whatsoever is able to fix the meaning of a concept. All they can do is constrain it in specific contexts. As Chang puts: "measurement operations provide only one specific context in which a concept is used, operational definitions can only cover one particular aspect of meaning" (Chang 2009). But the meaning constrained by operational definitions seems to be too narrow for even in scientific discourses: hardly any language users will agree that the definitions $length_1$, $length_2$, $length_3$, and so forth, i.e. the definitions based on the currently available procedures will exhaust the meaning of length. In fact, operational definitions are neither necessary nor sufficient for meaning.

In addition, one can also raise issues about the exact nature and the publicity of the operations themselves. One may rightfully ask what qualifies as an operation and what not. And why? Bridgman himself sketched a classification, where mental and "paper-and-pencil" operations also count. But this account encompasses so much that it flirts with triviality. On the other hand, qualifying laboratory measurements only is way too restrictive (Margenau 1956, 39). As Chang notes, a much more fine-grained analysis of operations would be needed (Chang 2009).

Not unrelated, we may ask how to understand a result of measurement. Can it be verified or checked? As it turned out, long after his major work had became seminal (Bridgman 1927), he was a desperate methodological individualist (many even accused him of solipsism). He maintained that in "checking and judging," as part of the operations constituting meaning and knowledge, a scientist cannot rely on anyone else but himself. Bridgman in his pressure experiments can only rely on Bridgman: the operations are private. With this he seriously deviated from logical positivism with its relentless quest for the ideal of protocol sentences, conveying personal observations to the highest possible objective availability.[21]

However justified are the above critical observations, operationism "fertilized" and "liberated" measurement in psychology, as chapter 3 will show. Nevertheless, I would be reluctant to call it a success story. On the other hand, operationalism has strong points when it comes to laying down valid concepts and procedures of measurement, as it will become clear on the following pages.

NOTES

1. This classification of concepts, which follows Carnap, is not at all pertinent or black and white (Carnap 1950, Carnap 1966). For instance, one can make the nagging note that by judging that this rod is one meter long one is just saying that this rod belongs to the class of one-meter-long objects (Kyburg 1984). Still I think it fits for setting up an initial scene.

2. One may find this definition too narrow, and insist that we should not exclude mathematical objects other than numbers. What is more, to take a more extreme view, any symbol may go. Likewise, somebody else could say that this throwaway definition is too wide, as we are prone to assign numbers to, say, washing machine programs, which hardly qualifies as measurement. I will address these points later.

3. For Frege, *concept* is a strict, well-defined concept, but we need not go into details here.

4. Following the ubiquitous practice, I use "iff" as an abbreviation of "if and only if" here and in what follows.

5. For an introductory text on Frege's work on the foundations of arithmetic and on the role of Hume's principle, see Zalta's article in the *Stanford Encyclopedia of Philosophy* (Zalta 2017).

6. As known, in this he did not succeed, nevertheless the merits of his work can hardly be overestimated.

7. Campbell defined measurement several ways. Karel Berka compiled the following further definitions of him (Berka 1983, 21): (measurement is) "the process of assigning numbers to represent qualities," "the assignment of numerals to represent properties according to scientific laws," "the assignment of numerals to things so as to represent facts or conventions about them."

8. Another method for measuring hardness is Brinell's test. Here we measure the penetration of ball of given diameter made of a given material, pushed with a given mass. It can be argued, that this method measures a *different property* than does Mohs'.

9. These rules have been living in many forms and wordings. Here I will follow Carnap because of his explicit treatment of empirical procedures as markedly distinct from mathematical manipulations. Even if his lecture-based book is a late-runner in its genre (Carnap 1966).

10. I have not mentioned the measurement of mass yet, I have always been talking about weights. But, of course, under given circumstances, measuring weight can be the base of that of mass.

11. New definitions are agreed to be introduced in 2019 based on the 2017 proposals of the International Committee for Weights and Measures (CIPM 2017).

12. It is to be noted that the prospective implementation of the new definitions changes the picture radically. One of the most interesting developments is that the definition of kelvin will depend on meter, second, and kilogram (ibid.).

13. For a detailed story of fix points and, in general, temperature measurement see Chang's book (Chang 2004).

14. To be fair I admit that my narrative here fails to account for some of Campbell's considerable contributions. Most importantly, in his Elements he develops an error theory as a prerequisite for establishing standard series in necessarily noisy, in particular, intransitive data (Campbell 1920). My only excuse is that a similar theory will be analyzed in the final chapter.

15. This is not to say that I am about to downplay the importance of "raw" observations. Quite the contrary, as it will be clear in these pages.

16. Interestingly, Bridgman also has something common with Michael Dummett's view on the difficulties around getting rid of our metaphysical habits. Bridgman held that to adopt the operationalist point of view requires immense change in our thinking and thus brings around major difficulties. Moreover, practicing this approach so diverged from the mainstream may result in social inconveniences even in a simple discussion on the state of affairs with a friend—well, it may indeed be annoying to demand an operationally firm meaning for every term—and may finally lead to isolation and misunderstanding. As Chang implies, he sort of foresaw his own fate in his later career as isolated and misunderstood in his thoughts (Chang 2009).

17. When one takes on the task of assessing Bridgman's operationalism, she has to face a special problem: it is not easy to tell what does it consist of exactly. Bridgman was not entirely systematic in his thoughts, maintaining somewhat incoherent claims at times. For some authors on operationalism a basic task is to reconstruct his achievements as a forceful system (Gillies 1972, 6–8, Chang 2004, 148–152, Chang 2009). Here I omit this issue, and give only a brief summary of the main objections against his views without assessing the soundness of the understanding they are based on.

18. We can mention Hanson just for one example (Hanson 1958).

19. "[The operationalist] fails to impart meaning to substantive concepts that is, concepts related to entities that are regarded as the carriers of operationally determinable qualities or quantities. To illustrate this latter point: it is possible to define, in terms of instrumental procedures, the charge, the mass, and the spin of an electron, but hardly the electron itself" (Margenau 1956).

20. See more on this in chapter 5.

21. Neurath's article is a graphic example (Neurath 1932).

Chapter 2

Rules, Procedures, Reality

The previous chapter was intended to provide a conceptual, rather than historical, introduction to the main issues this book is concerned with. This one pursues the problems to further depths and intends to make some original points, as well. I will address here questions like: how do we bring about units and how do we justify our ways? Is setting apart different kinds of quantities justified? Are the rules of their measurement genuine scientific laws? If yes, why? What is the role of conventions in what we call measurement?

Again, I will address the problem of realism regarding quantities. I will show that rational numbers are in every case enough to represent measurement data and that talking about continuous quantities (at least, in the widespread, classical sense of continuum) raises serious problems. After stating these preliminary results, I will take up the topic of realism again in the closing chapter.

2.1 TAMING TIME AND TEMPERATURE

Let us now recall the established rules of extensive measurement and see how they can be applied to length, one of the most mundane basic quantities. In line with the first rule, we are to have an empirical process to pick out equal lengths. We can lay any straight edges (for instance, rigid rods) side by side and judge whether they are equal or not. For now, forget about the problems of uncertainty by the limited sharp-sight or of the margins between which our assessments can be regarded reliable. If we lay two rods end by end along a straight line, the length so gained will be the sum of those of the two original rods—exactly as the second rule requires. Finally, we can pick a rod with a certain length, or can mark out an arbitrary distance on a straight edge, and regard it as the standard unit.

Having a straight wall of a certain length to measure, we can just begin measuring by first laying the unit rod by the wall so that the one end of the rod fits to one end of the wall. We mark the position of the other end, then move the rod by the wall so that the first end fits to the mark, and so on. When reaching the end of the wall so that the rod's end more or less fits the end of the wall, we count the times whenever the rod was in a new position by the wall, and then conclude that the length of the wall is *n* units.

Putting aside many suspicious presuppositions present in the above account, let us apply these rules now for time, another fundamental extensive quantity. This is a bit trickier. We may record an event *e*, and then record an event *f*, but how could we be sure that the so defined time interval equals to another one defined by events *g* and *h*, or not? Moreover, how can we assure additivity for these intervals?

The answer is that we can establish equality between time intervals spanned by simultaneous event pairs (Carnap 1966, 78–85). Similarly, additivity can be taken for any two adjoining time interval pairs, that is for any two intervals specified by events *e* and *f*, and events *f* and *g* respectively. Clearly, these concepts are much weaker than those of length in that we can establish equality or account for additivity only in naturally occurring, readily found situations. We have no ways for manipulation, we cannot carry over a time interval to compare it to others. Worse, we should also be clear on what is meant by simultaneity, which is not trivial in case of faraway events.

Anyhow, to make any use of it, the notion of time measurement must be strengthened, and this can be done by exploiting periodic phenomena. Traditionally, the cycle of seasons, the variation of days and nights, the apparent trajectory of the sun provided natural base for measuring time. Later this role was taken over by human artifacts with cyclic motion: pendulums and clocks. But no cyclic phenomenon is enough in itself. It is worth making explicit at this point that the procedure of time measurement consists of two components. For one, we need a dedicated periodical phenomenon. For the other, we need the "linear" process of counting. That is, we not only need to distinguish between individual cycles, but also to establish a consecutive order for them. This shows nicely how counting is integrated in more complex measurement procedures.

But how can we be sure that the chosen reference periods are really stable? We know that there are notoriously unreliable reoccurring events, such as the first snow in the year, my arriving at the office in the morning, or even my heartbeat. The Earth's motion or a pendulum are widely regarded as more reliable, but usually with some caution. We *know* that in practice the pendulum is slowed down by friction, and strictly speaking the Earth's periods around the sun or around her own axis are also not constant. Now, these convictions rely on some complex *theoretical* machinery of science, thus go far beyond the immediate *empirical* task of taming time. Worse, they presuppose an already established concept of time. The same is true about some periodic processes on the atomic level widely considered to have the most reliability, which nowadays serve not only as bases for modern chronometers, but also as conceptual foundations for theoretic issues of measurement. A commitment for a "right" concept of time is already there when we talk about their reliability.

What justifies the establishment of relative reliability between different periodic phenomena? Consider two pendulums, one with "normal" periods, the other with a "hectic" motion. We would, of course, take the pendulum with even periods as more reliable than the other. That is, the one we regard more stable intuitively. Unfortunately, intuition itself is often a quite unreliable tool—according to Poincaré it is in fact an illusion that we have any intuition on time spans (Poincaré 1905, 35–36). It can easily be imagined that the two pendulums are moving so that we have no clue which is the one with the "hectic" motion, we can lay down only that they are moving differently. It might be tempting to say that let us *measure* which pendulum is the one with the "hectic" motion, but it is easy to see that now we are in a bad cycle. For, in order to measure this, we already should be in possession of a "reliable" pendulum, and so forth. What is more, there is not much difference between the cases of length and time. Just as we may only suppose or stipulate that the duration of one swing of a "stable" pendulum is the same as the other, so can we only assume that our standard unit rod remains the same when being carried over in space (and time).

If we cannot have any clue on a "stable" unit, or even on relative reliability, we can take any periodic phenomenon we like as standard, any rod or any swing of any pendulum. There is no compelling reason to choose a specific one, logically any choice is equally justified. Moreover, we may arrive at a consistent description of the world based on any choice. As Poincaré maintained, we may choose Euclidean geometry or some of the non-Euclidean geometries in our descriptions of the world, just as we like, we can as well arrive at consistent theories, only the laws in our theories will be different (Poincaré 1902). For instance, we may describe relativistic phenomena by insisting that space is Euclidean. In this case we must conclude that our measuring rods expand and contract, and also our clocks run faster and slower. On the other hand, we may regard our readings of our chronometer as time and the unit rod as the unit length, and in this case we must conclude that space-time is non-Euclidean. Which one of these options is actually chosen is a matter of convention. There is no empirical test to decide which way to go, as both of the ways can be equally consistent with our observations. But after all scientists place their bets, and they prefer a choice resulting in less complicated laws, neater theories. The applied rules are chosen not because of their truth, but because they are convenient. They are "the fruit of an unconscious opportunism" of the scientist (Poincaré 1902, 36).[1]

Opportunistic or not, Carnap suggests a possible reason for preferring a periodical phenomenon over the other when measuring time (Carnap 1966). He insists that we can establish relatively stable behavior for a large class of phenomena and not for others. By relatively I mean that their behavior is stable relative to each other. My pendulum executes roughly equally many swings during each period the Earth turns around. What is more, I have a lot of other gadgets exhibiting the same stability: the hands of my clock takes more or less exactly as many turns each time my pendulum takes one thousand periods, and so on. It is not that we would assume some divine harmony here, we regard these coincidences as contingent facts, but we can bet on this large class, because we do not have such an extent class for other behaviors at hand.

Let me remark that the fact that other such large classes are not rightly available for us does not guarantee that they do not exist, meaning that they cannot be in principle established. But what is more important, the concept of relative stability is relative itself, in particular, it is relative to a degree of exactness or resolution. If we are just "zooming in," jumping a magnitude order, we will find that many phenomena are now dropped out of our favored class: in face of the new standards of exactness their behavior will not be found stable relative to the others in the remaining class. And going along this way our large class dissolves into smaller and smaller classes until we end up with some class with one (type of) element, for example, the class with the behavior of cesium-133.[2] Still, at a reasonable level of "accuracy," we can build our standards on phenomena exhibiting congruent behavior. This observation will be of great importance below.

At this point it is already clear how we can carry over time intervals as units in line with our established rules. We can take any periodic phenomenon as standard, for instance, the motion of our favorite pendulum. Take our third rule. At an arbitrary moment $t(e_0)$ we begin to count the swings of the pendulum. To any event e we can assign the (approximate) number of swings up to that moment $t(e)$. Now if $t(e)$ and $t(f)$ differ in n swings, and $t(g)$ and $t(h)$ also differ in n swings (be the two event pairs however remote), we regard the two differences equal. Still, we may have some doubts whether this account is satisfactory, and I will come back to the issue.

But before that, let us try to answer two questions brought up in the previous chapter. The first asked whether we are free in choosing our reference material which we base temperature measurement on. The second asked whether the scales produced by different materials are similar in the sense that they can be linearly transformed into each other. Now, the short respective answers are: yes, in principle, we are entirely free in choosing our material for temperature measurement; and no, different materials do not behave the same way and do not produce similar scales.

Consider figure 2.1. In the first scenario we choose material *a* as reference, whose expansion between our two dedicated temperature points is *l*(*a*). For the sake of simplicity, let us divide our unit into ten parts. Observing then, with the help of our established scale, material *b* with the expansion *l*(*b*) between the dedicated temperature points, so that we mark the respective degree marks as it expands when heated (say, we heat liquid thermometers in the same pot of water, and we are registering the liquid level on the glass with a marker), we may well find that the correspondence is far not linear. Similarly, in the second scenario, when choosing *b* as standard, we will find that material *a* seriously deviates from a linear expansion when heated.[3]

Now do we have any reason to prefer any of the possible scales? One may insist that there must be certain theoretical considerations and scientific laws determining the "right" interpretation of temperature. For instance, she could mention that the pressure of a certain gas in a closed vessel is proportional to the temperature. Postulating that it is linearly proportional, we can use this phenomenon to *calibrate* temperature scale. But if we use the so gained scale to observe another phenomenon, we may well find now temperature is not related in a linear mode to another magnitude (e.g., to the volume expansion of another material). How should we choose the "right" phenomenon to calibrate with?

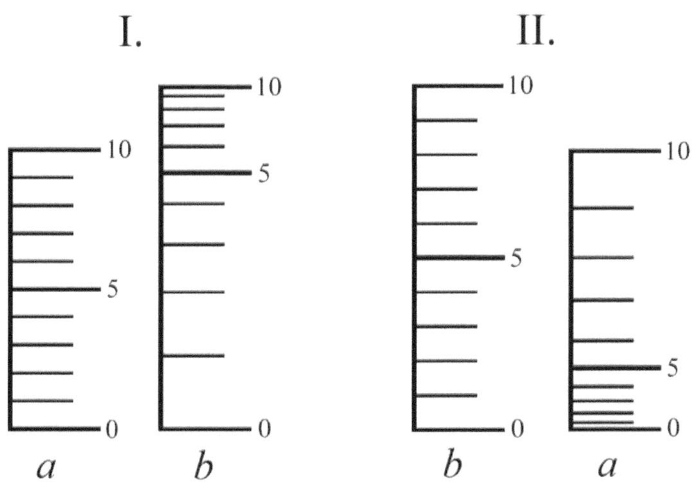

Figure 2.1 Relative Temperature Scales by Different Materials.

One may also say that the modern concept of temperature is based on the kinetic energy of particles. Cannot these energy levels define the "right" scale? The thing is, here temperature is simply *defined as* the kinetic energy of particles, that is, energy is just another way of talking about the same thing, but cannot help us solve the question of calibration. If we could measure and consistently relate the amplitudes of the oscillation of the particles in different materials to a common additive scale, *and* we could derive the procedure of temperature measurement from that of kinetic energy, we could say we have "extensified" the quantity of temperature. But this is not the case. The theories of statistical mechanics do not bring us closer to the *operalization* of temperature.

Serious problems arise also when trying to extend the scale we had based on a given phenomenon, such as the expansion of some material. First, there is no guarantee whatsoever that the behavior of the given material will (in any relevant sense) be similar to the one observed between the dedicated points beyond the boundaries of the interval. Second, every phenomenon is apt for measurement within narrow limits only, in other words, every material exhibits certain not so preferable behavior around its phase changing points.

This latter consideration does provide some reason for choosing an apt phenomenon as a base. Kelvin, the modern standard unit for temperature, is defined, roughly, as follows. Theoretical results of thermodynamics indicate an ideal, absolute zero point (in the terms of statistical mechanics, the one where no microscopic motion of particles is detectable). Take this one as the first dedicated point according to rule (iii). Take the so-called *triple point*[4] (very roughly the freezing point) of water as the second dedicated point. Then divide the interval so gained into 273.16 equal parts, this guarantees that our new unit will be equal to the one of the Celsius scale—there are many advantages in respecting traditions. But now, how to extend the scale upward? There are several ways, but it worth choosing a phenomenon that sweeps an immense interval. Take for example the ideal three-dimensional black body and take the characteristic frequency of black-body radiation as directly proportional to the temperature. It maybe a good choice because of the stable wide-range behavior, but it is still an arbitrary choice without any logical constraint. Moreover, it is surely unpleasant that we can never directly observe an ideal black body.[5]

Few would question that the measurement of temperature (similar to other quantities) is conventional in the sense that units and divisions of scales are determined by the consent of scientific community. Now it is clear that convention plays a greater role here: the characteristics of the whole scale may be determined by the choice of the defining phenomenon. But if the choice is entirely arbitrary and unmotivated, it raises serious questions of validity and uniformity, and after all, on the very meaning of temperature. Sticking to an arbitrary convention may justify assertions relating measured temperatures at different locations in the terms of colder and warmer, but it does not enable relating, for example, diurnal temperature variations. One could, of course, bite the bullet and arrive at the conclusion that, in spite of our ubiquitous practice, it is simply meaningless to talk about equal differences in different ranges of a temperature scale. Without a clear, valid procedure, rule (v) is idle and useless.

A way out from this frustrating situation may come into view by observing that the issue of temperature show some spectacular similarity to the case of time. We have seen that commenting on Poincaré, Carnap proposed to base our standard choice on phenomena which exhibit mutually stable behavior. Even if this commitment will not downright validate our ways, it will provide them with some empirical weight. This authority may be enough for building up procedures to align with the rule for equal differences. And if we have calibrated clocks by this principle, cannot the same be done with thermometers? It can, and it was, indeed.

2.2 RULES AND LAWS

But first, let us tarry for a while with the role and status of measurement rules. What kind of formulas are they? It is telling that Campbell prefers the term *laws* (of measurement) instead of rules, however, he usually seems to be quite cautious with verdicts on the nature of these laws. At some place though, Campbell explicitly defends his choice by emphasizing that the operative rules for measurement are genuine scientific laws. "They certainly are laws in their application to . . . properties such as weight or length. The fact that the rules are true can be, and must be, determined by experiment, in the same way as the fact that any other rules are true" (Campbell 1921, 119).

We may observe that, typically, a scientific law has the following form: $Q = \Phi(R,S...T)$, where Q, R, S, T, and so on are for different quantities, and Φ is some mathematical formula with quantity symbols as variables. For example, Ohm's law is usually put in the following form: $I = \frac{V}{R}$; and states roughly that the current on a given conductor equals to the quotient of the potential difference and resistance. True, certain other kinds of scientific statements are also often traded as laws, but the example sketched here is characteristic to physics—the discipline Campbell surely had in mind above all.

Even at first glance, the rules for extensive measurement are different in (at least) two respects. First, two of the three are naturally expressed as conditionals rather than equations. Of course, we can force ourselves to reword laws as Ohm's as conditionals. But their natural equational form suggest that they hold, or at least are supposed to hold, *universally*,[6] whereas it seems that the rules of measurement may hold under one set of circumstances, but not under the other. Sure, not all qualities are measurable. True, one could object: all right, but when they are measurable, they are universally measurable. But this amounts to saying that once we have an empirical procedure for measuring a quantity we will always have it (or have one) under whatever circumstances. It is rather doubtful, I suppose. Just consider the case of mass measurement by a balance scale through weights. Our rules would probably be useless on the International Space Station.

Second, while laws of physics usually involve a great deal of quantities, the very rules of measurement remain entirely silent on them. These rules do not say anything about single quantities, but provide some empirical principles for rightfully establishing a large class of them. Thus, they may be regarded as methodological guidelines, or test requirements—that is, some kind of meta-laws.

Nevertheless, I can easily agree with Campbell that the empirical procedures satisfying these rules of measurement involve experiments. Thus, they involve a great deal of sophisticated equipment and operations. Sadly enough, our rules remain utterly silent on these details. It is to be noted that having

Figure 2.2 A Good Start toward Empirical Equality no. 1.

the "right" tools for a given experiment is not unproblematic in general. But while I am aware that in case of an electric experiment I will have to begin with some voltage sources, resistors, capacitors, and other stuff of this family, I have no clue whatsoever how to deal with extensive quantities as such. What is clear is that no unified toolkit exists for them.

Anyhow, Campbell seems to suggest that once we found the appropriate equipment, any difficulties can be overcome by the experimental method itself:[7]

> [F]or instance, [a rule] is only true if the balance is a good one, and has arms of equal length and pans of equal weight. If the arms were unequal, the rule would not be found to be true unless it were carefully prescribed in which pan the bodies were placed during the judgment of equality. Again, the rules would not be true of the property length, unless the rods were straight and were rigid. In implying that the balance is good, and the rods straight and rigid, we have implied definite laws which must be true if the properties are to be measurable, namely that it is possible to make a perfect balance, and that there are rods which are straight and rigid. These are experimental laws; they could not be known apart from definite experiment and observation of the external world; they are not self-evident. (ibid., 119)

Figure 2.3 A Good Start toward Empirical Equality no. 2.

But how do we know that we have a (near) “perfect” balance scale? Surely, it is not only stipulated that a good balance scale exists, since it could not be known apart from “observation of the external world.” One must, then, have the experimental means for determining that the balance scale which we use for combining weights is a “right” one. But to determine this, it seems, one must already have the concept of equality for weight, the very concept one approaches through the rules for weight measurement. Since the appropriate experiment would be nothing else than *measuring* that the scale is unbiased by equal weight in its pans.

In fact, it is not even enough to conclude that the scale is a good one. Suppose we have two weights of which we, let us allow it, do not *know* whether they are equal in mass, but we suppose so, because *apparently* they are. Having them in different pans and observing balance is a good start. But when exchanging the two weights we observe a bias, we can rightly conclude that something is wrong. When observing balance, we need one more piece of experiment: we must put the two weights in the same pan. Finding that the very pan has sunk provides some reason for believing in our scale, and repeating this piece of experiment in the other pan with the same result provide further grounds. (Strictly speaking, it is still not enough to know that we have a good scale. It may well be that it is only insensitive to the difference between the two masses.) But we can arrive to this conclusion only because we presuppose that mass is measurable (or more exactly, that individual weights are totally ordered).

Notice further that one and the very same procedure applies for the cases where the masses are apparently different, or the scale is apparently biased or both. If we observe equality for the weights in different pans in both

Figure 2.4 A Good Start toward Empirical Equality no. 3.

possibilities, and we observe the sinking of a pan when both of the weights are placed in, *and* we have some pairs of weights of which we cannot establish equality this way, we must conclude that we have an empirical operation for equality independently of our impressions of appearances.

I will not examine the case for the other rules. It is enough to see that their test requires identical (or "very, very similar") copies of a unit weight, which can be arrived at by equality tests also.

All in all, do we have an operation also to pick the "right" balance scales without the measurability presumption? It remains to be seen that we cannot claim, for example, the equality of the balance arms. We may say that we will find that a "right" scale has equal arms *in most of the cases*, but this assertion relies on some vague theoretical assumptions on torque, rather than on actual observations. We may as well assume some peculiar distributions of friction, or even complex spring mechanisms inside the tool.

As stated above, the rules of measurement differ from physical laws in important respects. I am inclined to say that they are methodological considerations on the empirical operations underlying measurement procedures. Roughly, they suggest we *should* have such and such empirical procedures in some way similar to mathematical operations in order to assert that we are dealing with a measurable quantity. Whether we are smart enough to find such operations by carefully assembling our gadgets is a contingent fact which neither confirms nor disconfirms the rules by itself. By that fact either we can assert the given property's measurability or not. But honestly, when we are in trouble with measuring, say, a faraway star's core temperature, the last thing we would conclude that temperature *there* is not measurable. For sure, we will have some theories.

Figure 2.5 A Good Start toward Empirical Equality no. 4.

One of the clearest empiricist accounts on the status of measurement rules I found is given by Arthur Pap, who seemed to build a footbridge from Carnap to Quine (Pap 1959). To him (and Carnap), these rules belong to a larger class called rules of *correspondence*. As the name suggests, these rules provide links from laws and theories to observational data. As such, they are outright incapable for confirmation or falsification, whereas genuine scientific laws might be falsified—in theory. But this is not the whole story. Quantities obtain their meanings from the laws and theories in which they are present on the one hand, and from the rules of correspondence on the other. Properly described procedures work as operational definitions, and as such *partial interpretations* for quantities. None of the interdependent quantities present in a law is fully interpreted, consequently there are only partially interpreted systems.

According to Pap, this entails that "it is hopeless to try to reconstruct a quantitative scientific theory dualistically as a system of statements some of which are analytic . . . and some of which have factual content" (ibid., 187). The analytic-synthetic distinction fails. Being so, no observation is ever able to confirm or falsify a single statement, only the system as a whole can be the subject of revision under sufficient reasons. Not in the least, this way the rules of measurement are neatly "hidden" in the theoretical system.

2.3 CONCATENATION

One may still insist that there must be certain requirements warranting the validity of a chosen measurement procedure. Talking about extensive quantities, some authors (most notably Campbell) suggest that such a requirement is additivity: empirical addition of two magnitudes must yield a number which is the result of the arithmetic addition of their respective measured values. By contrast, Carnap argues that it is better to speak of an extensive quantity when we can think of a *natural* concatenation operation for it (Carnap 1966). Thus we are able to speak of extensive, non-additive quantities. Let us take the addition of velocities in special relativity theory. Here, concatenated velocities are not summed up by the addition operation of arithmetic, but by an equation given by Lorenz-transformation:

$$v = \frac{v_1 + v_2}{1 + \left(\frac{v_1 v_2}{c^2}\right)}.$$

Likewise, to give a less intricate example, we may like to see the measurement of angles as extensive. But angular addition differs from the arithmetical one in that we do not have a value greater than 360. Thus, when measuring angles, 270 + 120 = 30. So, with Carnap, we can embrace the following strategy: while insisting on a "natural" concatenation concept for extensive magnitudes, we go beyond the notion of additivity to encompass other operations beyond mere arithmetical addition.

If we were already puzzled in choosing a proper concatenation operation by rules, now we may be more perplexed. Take length, where we supposed to lay rods end by and along a straight line. Not only that defining a straight line or geodesic before framing the concept of distance is a bit problematic, but the very nature of the operation is unconstrained by the extensive measurement rules, let alone when we drop arithmetic additivity. Thus, so to say, any consistent operation goes. We may as well take laying the rods by the legs of a right angle as the proper concatenation operation. Of course, "addition" in this case will be a "right angle addition" (Ellis 1966, 80), one following the Pythagorean theorem. The point is that the operation so defined will satisfy all of our rules of extensive, fundamental measurement. The choice of the proper concatenation seems to be, to a large extent, conventional. Still we can say that as long as we have a meaningful and "natural" concatenation at hand linked to a consistent manipulation of numbers, we can legitimately talk about extensive measurement.

But what is to be "natural" enough to qualify as concatenation? A skeptic about a meaningful demarcation, Kyburg argues that the existence of a proper concatenation operation is a question of taste (Kyburg 1984). Take the archetypal example of intensive (non-extensive) quantities, temperature. Temperature is widely held not to be additive: having a body with a temperature T_1, and another one with a temperature T_2, there is no way to combine them so to have a body with $T_3 = T_1 + T_2$. Worse, we have no concatenation operation for temperature of any kind, it is often said.[8] Well, we do have, says Kyburg, just take "putting the two bodies in an oven and heating them until T_3" as the proper concatenation operation (ibid., 17). One may say that this notion is unnatural, but no one can insist that there is a strict, a priori border between natural and unnatural, so the distinction of extensive and intensive magnitudes is a matter of degree and taste, Kyburg concludes.

I willingly let it. But I would not stop at this point, a closer analysis of this construction bears more important morals: the unattractive peculiarities of the above example turn out to be essential features of a general concatenation concept. To see this, let us try to image how one would argue against regarding Kyburg's example as genuine concatenation.

First, he could exhibit a clear circularity in Kyburg's "temperature concatenation": the operation requires an already established temperature concept. In order to heat the oven to a given temperature, we must have a temperature scale at hand. But the main thing to do with concatenation is to establish the unit and thus the scale. If we are free to heat the oven as we like, the concatenation of T_1 and T_2 would yield any temperature. To sum up: Kyburg's notion of concatenation for temperature is either void or circular.

Second, one could make a case on the persistence of prototypes. When heating the two bodies in the story up to T_3, we will not have two bodies with the respective temperatures of T_1 and T_2 anymore, but we will have them with the temperatures of T_3 and T_3 (respectively). So their concatenation should result in a body with a temperature of $T_4 = T_3 + T_3$, but for this we should again use the oven-based concatenation operation—and so on infinitely. That is, we can never have our initial magnitudes and their "concatenation" at hand simultaneously.

Big deal, Kyburg could shrug, this is just some time index jugglery, and anyhow, where is it stated that concatenated magnitudes should be available after concatenation? Further, it may easily happen that relying on these indices the disputant can build a definition of a "proper" concatenation matching his own taste. One that, for instance, gives account for the availability or unavailability of intitial magnitudes or prototypes. Again, no rational reason can be given for disqualifying such cases of concatenation, unless we embrace some ad-hoc monster-barring strategy (Lakatos 1976).

Monster or not, the above concatenation concept seems to encapsulate the essence of all concatenation concepts. True, when one establishes scales through concatenation, it is usually presupposed that a prototype is always there for the next step in the procedure. The standard measuring rod of length L, or one of its identical copies, can always be laid end-to-end to the last copy or last mark by the straight line on which the procedure is being carried out (along with the rules). But the only reason for using standards (or prototypes) is that it is taken for granted that they remain the same from one moment to the other, or taken over in space from one place to another. Otherwise we could make any choice, we could just take any rod of whatever length. But is there any reason to believe that the length of the measuring rod stays constant the whole afternoon? More generally, is it justified to suppose that $L_{t_1} = L_{t_2}$?[9] I would not say so at all. In order to empirically justify it, we should carry out measurements on its length at different times. But our concept of length is based on this very concatenation concept. Recall, further, that classically length measurement is conceived as a manipulation of a standard rod along a straight line. Now, what is meant by a straight line? Without a doubt, an obvious choice is to conceive it as picking the shortest path between two points. But, alas, we are about to establish the very concept of length.

Historically, non-monster concepts of concatenation feature an important class of measurement procedures. The importance of this class lies in the belief that the numerical representation of the quantities at issue is immediately (and intuitively) justified by the procedure itself. Now we can see that wholesale concatenation concepts suffer from the defect of circularities, and this fact casts doubts on their fundamental role in measurement theories. Neither they are able to set apart basic concepts of quantities, nor they are lucid and robust enough to provide empirical foundations for measurement.[10]

2.4 CONVENTIONS AND CONGRUENCE

With the traditional importance of concatenation in the conceptual development of measurement in mind, abandoning it is disheartening: what remains to build on? In this section, touching on some recent studies in the topic, I roughly indicate the direction of a possible way out. Further details will be given later in this book.

To begin with, we need to get rid of all of our tacit presuppositions as potential sources of circularities and clearly identify the conventional elements in a potential candidate for a cogent account. László E. Szabó provides an operational approach to space-time along these lines (Szabó 2009). He gives an empirical definition for space-time tags in a single inertia frame, relying on a single standard clock. We do not presume much of the etalon clock: neither that it reads some "proper" time, nor that it ticks evenly. We require only to have the clock's readings whenever we need them. But a lonesome clock floating alone in the universe is not enough for tagging, let alone coordination. We also need further gadgets: the so-called *markers*. Our expectations toward them are less modest: they must be able to be triggered by physical events, and to receive, cast, and recast light (or radio) signals. We stipulate a marker at the standard clock, continuously broadcasting the clock readings. Each marker must have a unique ID, and whenever a radio signal leaves a marker it must contain the ID and the clock reading. We suppose to have as many markers as we need.

With this machinery at hand, we define the time tag of an event e as follows:

$$\tau(e) =_{df} t_1 + \varepsilon\,(t_2 - t_1),$$

where t_1 is the clock reading at the departure of a signal from the standard clock, and t_2 is the reading at the arrival of the signal reflected by a marker at event e.

Let us sum up the conventional elements of this setup at this point. No doubt, we chose our standard clock by mere convention, but it also must be clear that to rely on radio signals is also conventional: to explain our choice in spatio-temporal terms would be circular. Also, the "direction of time," or the order of the clock readings, and even causal order—by the "sending" and "receiving" event types—conventionally given. Finally, in the equation above, $\varepsilon = \frac{1}{2}$ by convention.

We can define distance from the standard clock by $\frac{1}{2}(t_2 - t_1)c$, where (by convention) we can give c the value traditionally assigned by the scientific community to the velocity of light—whatever this latter notion means. But this is not enough to set up a system of unique space-tags: we have to define absolute distance as well. In order to do this, we need the concept of *rest time sequence*: a unique time sequence tied to a given locus. I will not go into the technical details here. What is important to stress is that it is an empirical question whether rest time sequences exist or not. If we are lucky enough, we will find that there is a unique rest time sequence for every event, picking out a "world line" of a possible object which is at rest relative to the standard clock. (What this "fortune" means for us will become clear in a moment.)

Again, with this approach, the whole system of Euclidean geometry is stipulated as a class of empirical facts! And with these facts in hand, we can provide a spatial coordination in a three-dimension frame. (Technical details are again omitted.) Now that we are justified to use spatio-temporal terms, we are able to talk about inertial motion, which will be a time-like straight line in the usual four-dimensional Minkowski-space.[11]

Now for the dessert: Szabó provides two interesting results, each given as an output of computer calculations:

(i) There is no rest time sequence if the standard clock moves non-inertially.

(ii) There is no rest time sequence for every event if $\varepsilon \neq \frac{1}{2}$.

Practically and sloppily put, this means that the predictions of our usual theory on space-time can be true only if we have an evenly ticking, non-accelerating standard clock and our conventional choice on a given constant is appropriate. That is, we have a complex theory which implies the existence of an instrument with such and such properties. Conversely, given such an instrument, our incumbent theory on space-time can be tested, at least, in principle:

> Whether these statements are true or not is, therefore, an empirical question, and it is far from obvious whether they would be completely confirmed if the corresponding experiments were performed with higher precision, similar to the recent GPS measurements, especially for larger distances. (ibid.)

Let us not meddle now with the problem of complete or sufficient confirmation, we have another issue here which is, though not entirely unrelated, more important for our recent concerns. Namely: how to understand "with higher precision"? To take it to mean "more exactly" is hardly a wise choice: not the least, circularity is creeping back through the backdoor. For we can regard our equipment exact iff the predictions of our theory are true. To assess how exact our gadget is is nothing else than to *measure* how much its readings deviate from the "real" values. Let it: we have to live with approximations, yet the question remains, how to measure our approximations so to order them?

No doubt, the above interpretation flirts with metaphysical realism: it sets "the world as it is" against our relentless trials in more and more approaching its truths. Szabó would hardly be happy with such an account (nor would I, as it matters). It would be much better to talk about (something like) precision without giving any inkling of such metaphysical commitments.

In fact, our conventional standard clock is not just there floating around without any means to address its reliability: its stable behavior is *brought about* by tiresome human activity. In a previous section with a passing mention I incriminated cesium-133 as the final base of our contemporary time standardization. Now, as Eran Tal witnesses (Tal 2016), there is much more to this story: *timekeeping* is a highly complex human endeavor. Let us see it in some detail. Indeed, a handful of cesium fountains[12] provide what is called *primary standard*, which is used for adjusting atomic clocks exhibiting *secondary standard* (a somewhat more populous class). These latter gadgets are running continuously, and are very stable in the short run, but sooner or later diverge from each other and need to be adjusted by primary standards, such as by the "readings" of the cesium fountains, which are in work for short terms and only several occasions a year.

At this point a good deal of theoretical and institutional machinery enters the story. To approach the ideal of clock readings all around the rotating surface of the Earth (called terrestrial time), an operationally manageable measure is introduced. This is called Coordinated Universal Time, or UTC. As a first step, the readings of a few hundred atomic clocks realizing secondary standard all around the world are to be aggregated into a scale known as free-running time, or EAL after the French phrase. The method of aggregation is of great interest for our resent investigations; here I quote Tal's account verbatim:

> EAL is an average of clock indications weighted by frequency stability. Finding out which clocks are more stable than others requires some higher standard of stability against which clocks would be compared, but arriving at such a standard is the very goal of the calculation. For this reason EAL itself is used as the standard of stability for the clocks contributing to it. Every month, the BIPM [Bureau International des Poids et Mesures (International Bureau of Weights

> and Measures in Paris)] rates the weight of each clock depending on how well it predicted the weighted average of the EAL clock ensemble in the past twelve months. The updated weight is then used to average clock data in the next cycle of calculation. This method promotes clocks that are stable relative to each other, while clocks whose stability relative to the overall average falls below a fixed threshold are given a weight of zero, i.e. removed from that month's calculation. The average is then recalculated based on the remaining clocks. (ibid.)

To have the whole picture, we have to mention some further facts. Practically all of the participating atomic clocks are built by one manufacturer, and shown to be quite stable relative to each other. On the other hand, the worldwide list of participating national laboratories are highly affected by contingent issues of diplomacy. Again, comparison of the remote clocks is realized by the GPS system, thus signal transfer is exposed to atmospheric conditions, and possible discrepancies in GPS time (also derived from UTC), *time transfer noise*. Apart from theses contingencies and sources of uncertainty, we need elaborated algorithms to cope with other factors too. We need a weighting algorithm for aggregating the readings of free-running clocks which excludes overly influential clocks and cushions frequency jumps. Further, we need some "steering" measures, because secondary standard has an observed tendency to drift away from the primary standard, the first being slightly "faster." These algorithms are fairly unconstrained and run with largely arbitrary parameters—they are given by convention.

This is definitely an abridged edition of the story, still worthy of some crucial observations. We should, of course, address the apparent circularities and the role of ad hoc commitments. We should, again, examine the social, institutional, and perhaps even the instrumental aspects of timekeeping activity. Indeed, I will come back to some of these questions later. For the time being, a few more words on observational regularities, uniformities or, as I call it, *phenomenal congruence* are in place. It is clear that the current practice of timekeeping shows, extracts, squeezes out, or lives on empirical regularities, without relying on the problematic concepts of concatenation or exactness. And this is the way we may be able to find the "precise enough" clock for Szabó to test the prevailing space-time theories: with relentless and active quest for congruence.

To be sure, this sentiment is not without antecedent examples in the history of science. Take again the case of temperature: Hasok Chang vividly covers the story of the calibration of thermometers in his book (Chang 2004, 74–83). Regnault, a meticulous experimenter, did a heroic work in picking out the "best" thermometer possible. His approach was largely anti-theoretical, as based on the notion of *comparability* only. He managed to sort out a set of gas thermometers which was comparable, that is, provided similar enough readings under the same circumstances. In my wording: he managed to sort out a set of congruent phenomena. Note that he restrained himself from asserting that he established the "real" scale of temperature, and it would not have been justified to do so at all—even if his achievements later got some theoretical confirmation (ibid., 192–196).

To what extent was his approach really theory-independent and what tacit presuppositions he really maintained are interesting questions, which I will not address at this point. For now there is one important moral: even when having no explicit theories with successful operalizations at hand, we still have a choice to establish a firm quantitative concept with sophisticated manipulation or, so to say, empirical brute force. To put it in other words, through *epistemic iteration* (ibid.), we are justified in establishing phenomenal congruence. I will pursue this issue further in the last chapter of this book, when talking about the construction of valid procedures.

2.5 CONTINUUM AND REALISM

One may maintain a view that exhibiting standards is one thing, flesh and blood measurement is another, the two do not necessarily go hand in hand. The latter is traditionally seen as a class procedures manipulating standard units with which some quantitative properties of the objects of interest are estimated or approximated. Above we have seen that to establish these units is far not trivial. So let us now address the estimation part.

Approximations seem to be essential to measurement. To improve our ways, we divide our standard unit into sub-units, as meter is divided to 1,000 millimeters, and minute is divided into 60 seconds. But it is easy to see that doing so we only export the problem of exactness one level lower. We can measure the length of a piece of metal with our measuring rod with a millimeter scale on, the ends of it will match those lines on the scale only roughly. Of course, we can still make our scales and tools more precise (take a vernier) and we can jump another order—and so on and so forth.

As we improve by dividing our scales into more fine-grain scales, one could suspect that measurement can never arrive at anything else than rational numbers, in other words, we can never assign irrationals to measured features—even if we could refine our means *infinitely*. Irrationals are results of theoretical considerations and calculations, in practice, we can never arrive at π when measuring the ratio of the perimeter and the diameter of a piece of land of a circle shape. Of course, here we are not after the point that no perfect circles exist in reality. Rather, the aim is to inquire into a discrepancy between intuition and observation. While we maintain that most of the quantities are continuous, the very nature of our measurement procedures seems to be such that we can only exhibit rationals with them, ever.

In fact, in a trivial sense irrationals can be results of measurement: one can report them as such. Maybe we could reword and sharpen our supposition as: no relative irrationals can be justly reported as results of a given measurement procedure. But having a measuring rod with markings like, for example, $\sqrt{2}$, $\sqrt{3}$, and $\sqrt{5}$, nothing prevents me to report measured distances as $\sqrt{2}$m, $\sqrt{5}$m, $(\sqrt{5}-\sqrt{3})$m, and so forth *justly*, or according to my best knowledge. Thus, in what follows, I endorse a weaker assertion: rationals are *enough* to represent any measurement results. Below I will defend this thesis against some possible objections, then I explore some of its consequences with regard to the problem of realism. This opens the door to further important considerations exposed in later chapters of this book.

Let us begin with the following situation. Consider a clock with only one hand for the indication of hours, motored by any of our cherished periodic event types. The hand is extended as a straight line, like by a laser light beam. This way, besides the ones on the clock face, we will have readings also on the floor—on a plane perpendicular to the clock face. These readings will, of course, fall in a straight line on the plane. Let us denote the floor readings at four, five, six, seven, and eight o'clock as *a*, *b*, *c*, *d*, and *e* respectively; see figure 2.6.[13]

Now, most of us would agree that we are able to measure the time elapsed from four to five by the length the hand took on the edge of the clock face (or by the change in the angle between the hand and the six-twelve axis). But then we may also hold that we measured the laser beam's way on the floor, that is, the distance between *a* and *b* at the same time, whose value is up to the unit choice and the length of the perpendicular beam from the clock (the one at *c*), but it is *very likely* to be an irrational number with respect to the given unit. For instance, if we take the length covered by the beam at the

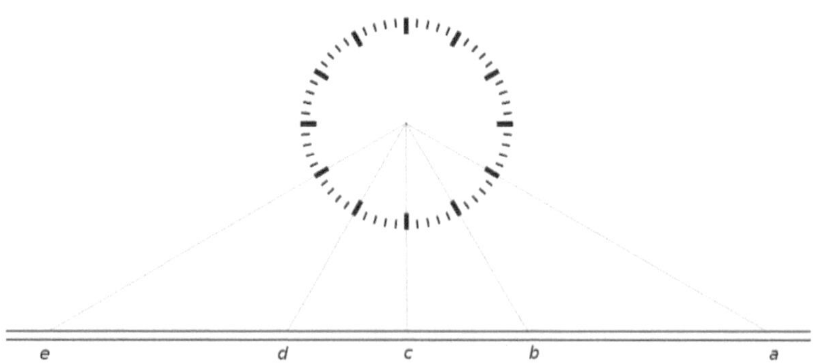

Figure 2.6 Clock Readings and Projections.

edge of the clock face by one hour, $r\frac{\pi}{6}$ (where r is the radius of the clock), and the distance l_0 between c and the center of the clock is $2r$ (i.e., $2 / \frac{\pi}{6}$, i.e., approximately 3.81971863427 . . . units), then by trigonometrical methods we can *calculate* that the progress of the laser beam from six to seven, l_1 is $\tan 30 = \frac{l_1}{l_0}$, approximately 2.20531558172 . . . units. Now we can state that we have *measured* the distance between c and d, and it is an irrational number. The strength of this argument lies in the observation that these numbers coincide not with some arbitrary measurement results, but with the limits of our standard units.

But let us take a closer look. What we have here is an idea of a function mapping all of our readings to some point on a straight line. It will map all of the readings from nine to three o'clock to infinity, but why not? It is not a one-to-one mapping, then. But let us now disregard the upper part of the clock.[14] One may assert that we have co-measured the laser beam's progress with time, but what we see here is a bunch of calculations. Are we to insist that we have also co-measured the actual *length* of the beam (e.g. l_0), just because we expressed it in the given units? For me it seems that these are trigonometrical calculations independent of any measurement. This is not at all to say that no measurement procedure should involve calculations. What I am up to is to indicate that here we have an issue of mere *renaming*. As measurement data is reported as numbers, I could take any function I like to rename pieces of data. Consider just replacing the readings on the clock face by the square roots of primes respectively, as: 1 to $\sqrt{2}$, 2 to $\sqrt{3}$, and so on. Are we really ready to accept that now we are "co-measuring" rationals and irrationals?[15]

Further, suppose there is an ideal point on the edge of the clock which, by the motion of the beam, exactly coincides with seven o'clock, or with one of the endpoints of the time span between six and seven. Now, however we are trying to make our readings more and more precise, there will always be some uncertainty as to when exactly we are at the required point (expressed in some tinier and tinier sub-units), that is, at seven. Now whatever irrational number is calculated at the floor-end of the laser beam, we can never ascertain that we have arrived at it. That is, our empirical methods will not be sharpened simply by renaming a rational to a given irrational number. In a realist wording: our reports on measurement results are always to be understood as complex statements on intervals (and probability).

In a different setting, we could redefine any of our unit x as $\sqrt{2}x$ from tomorrow. That would cause much confusion, but change nothing essential. Of course, our unit choice is conventional, and even every unit can be regarded as irrational in face of another unit. If I would measure the weight of the standard prototype of kilogram against a piece of clay I just nipped at random, almost surely I would find that by endlessly improving my measurement methods, I could only approach it with my "measurement generated rationals" (accepting the generous help of the robust armature of subunits and identical copies in the measurement operations), and this process never ends. This means that even if we are to devote ourselves to a view that mutual irrationals (in some sense) belong to reality, we can be sure that our operations can only approximate them.

Again, one may wonder much depends on how *reporting a result* is conceived. But this is not so. We can regard reporting as an error-prone act of reading instruments scaled with whatever symbols, or we can simply equate reporting with establishing a result of measurement. In any event, for any set of reports (any body of data) I can always offer a consistent renaming using only rationals—where by consistent I surely mean that the order of the values is preserved. But there is more to it. It is clear that the difference of two relative irrationals cannot be bridged *exactly* by rationals. But I can always offer values bridging this difference with an exactness well beyond the reliability of our measuring tools. Indeed, I can exceed it in *arbitrarily many* orders.

Note further the fact that a series of reporting, being finite, cannot even be dense.[16] This suggests that we do not even need the whole armature of rationals for measuring. In fact, finite decimals are well enough.

Interlude. To further exploit our thought experiment for the sake of an illustration to the problems sketched in the previous chapter, observe that the path of the beam on the floor *is* suitable for time measurement even if we are agnostic to the clock face far up in the sky. Of course, we know that the beam first passes *a*, and *later* it passes *c*. What is more, we know that meanwhile it passes *b*. Thus, we can measure the order of moments, and we can even set up a daily (or a twice a day) meeting by referring to a point on the floor.

What we do not know is that *b* is at "half-way" in time between *a* and *c*. At this point we cannot be sure what reading corresponds to the middle of the time span between *a* and *c*, but nothing prevents us to take b′ as the corresponding point so that b′ lies exactly in the middle of the straight line section between *a* and *c*. Of course, b′ will not coincide with *b*. Still, we are free to take the (a, b′) (or (b′,c)) section as our standard unit for time measurement. Thus, adjusting the right end of this unit section to some arbitrary *x* point on the path of the beam we will arrive to the notion of *one hour later*. And it will be a neat, operationally defined concept. Until we do not *calibrate* our system with other time measuring tools based on periodic phenomena regarded as standard by consent, we can rely on it as our favorite, best and only equipment.

Now, it is time to address the realist stance: what is the reason to believe that there are reals "out there?" Why should we regard most of the quantities as continuous? And why would abandoning this belief be a "revolutionary change" (Carnap 1966, 89)? We might even find this sentiment strange, since the assumption that space or time has a structure analogous to that of real numbers cannot be confirmed by direct observation, not even in principle—it is clear from the fact that rationals are well enough to represent any set of measurement data. So, is it not possible that all of the quantities are discrete after all?[17] Physical theories already exhibited several phenomena of discrete nature, for instance electromagnetic radiation carries energy in quanta, or charge also has a smallest possible magnitude. Not in the least, even to interpret length on an atomic level is problematic in itself. (There would be no problem in geometrically describing discrete space and time: we can consider points x and y as neighboring if no events could take place between them.)[18]

But to comment on the discrete space-time problem is not my issue here. My concern is why do we see quantities as continuous? One may mention our deep, instinctual realist convictions as a reason. We can venture it is somewhat "natural" to regard present as a continuously moving frontier between future and past—anyhow it is a widespread way of looking at time; as it has also been a popular stance in the history of philosophy to regard time intuition as the very base of the real number concept. Again, our intuition tells us that the bullet fired from a gun, having a 10-meter-long trajectory toward its target will pass the point at $\sqrt{2}$ meter. Through the realist's glasses (suppose her to be realist in a semantic sense),[19] the sentence asserting the above state of affairs will be true independently of our knowing it (by observation), and what is more, independently of our outright inability to verify it.[20] Moreover, a realist may insist that as the bullet touches every point on its trajectory, it makes sense to say that the bullet (or an infinitesimally small part of it) is at a rational (or an irrational) number away from its starting point taken in meters at a given time. In other words, she may insist that the assertion that the bullet (or an infinitesimally small part of it) is at a rational at t_x is true or false independently of our knowing which, and independently of our not being able to know it ever in principle. This situation can be nicely described by the notorious nowhere continuous Dirichlet function defined as $\varphi(x) = 1$ if x is a rational, $\varphi(x) = 0$ if x is an irrational, by naturally representing "true" as 1 and "false" as 0 (Davis and Hersh 1981, 264).

True, continuum is usually associated with much prettier functions. Another possible reason for sticking to a model of continuous quantities may be the essential role, which mathematical analysis plays in scientific theories. The presupposition of continuous change allows us to talk about not only function values at certain points but also about their arbitrarily small environments as well, so not only about state of affairs but also about trends. Thus we can derive a velocity function from a position function, an acceleration function from a velocity function. As a consequence, whole scientific theories are presented as a bunch of differential equations, as equations relating basic functions to their derivatives. So the scientists' adherence to the continuum is somewhat understandable. But the everywhere continuous and derivable functions we used to use in our physical descriptions of the world are not conceptually necessary at all. Neither the classical realist approach to the continuum nor the applied mathematics constrain this. We can easily imagine, or mathematically describe cases where a body is at x_1, y_1, z_1 at a moment, but it is at x_2, y_2, z_2 at once, "in no time." In this case, our motion function is "broken," or not continuous.

In order to show that our classical realist model even leads to outright conceptual impossibilities, Dummett introduces other, more sophisticated examples of discontinuity (Dummett 2000). At some point, for instance, he presents a version of Thomson's lamp.[21] Instead of a lamp and a switch, consider a pendulum with an accelerating motion between its endpoints *a* and *b* lying equally far from the center of the pendulum's path *c*. Now the pendulum swings first from *c* to *a* in $\frac{1}{3}$ minutes. Then from *a* to *c* in $\frac{1}{6}$ and then from *c* to *b* in $\frac{1}{10}$ minutes. We can calculate the *n*th quarter swing as

$$S_n = \frac{2}{(n+1)(n+2)}.$$

Now, what position will the pendulum hold after one minute? We cannot tell, of course, since the sequence converges to 1, thus the pendulum must take infinitely many swings by then. The mere fact, that such thought experiments can be worded makes the classical model more than suspicious, at least according to Dummett:

> The classical model is to be rejected, because it fails to provide any explanation of why what appears to intuition to be impossible should be impossible. It allows as possibilities what reason rules out, and leaves it to the contingent laws of physics to rule out what a good model of physical reality would not even be able to describe. (ibid., 505)

I would reconstruct Dummett's argument as follows. Scientific theories used to live on continuous, differentiable functions. In particular, scientific laws generally establish relationship between continuous quantities. Now, these properties are constrained by contingent "facts," not our inner "model of physical reality." Worse, the latter is fatally incriminated once caught red-handed in showing conceptual impossibilities possible. So let us choose a new model which excludes these conceptual calamities (e.g., by rendering all the functions continuous and derivable—see below). As for me, I cannot really see what should shape our "model of reality" if not contingent proceedings of physics. Nevertheless, if it is the case that we are somewhat free to choose our model, I agree to choose the more streamlined one with less inner discrepancies.[22]

After so setting the stage, Dummett offers an alternative to the classical view, so to get rid of the problematic metaphysical stance while trying to preserve the strength of our scientific theories (ibid.). He suggests to apply the principles of constructivism for our reflections on the physical world so to get rid of the conceptual discrepancies of the classical approach. The constructive mathematical practice is largely based on the ideas of intuitionism,[23] a school that originates in the work of the Dutch mathematician Jan Brouwer. The advocates of it maintain that the objects of mathematics neither are real entities in some Platonic realm, as the rival school of logicists often holds, nor are mere figures of ink on a sheet of paper manipulated by the rules of the game, as the so-called formalists insist, but constructions in the mathematician's mind (or better, in an ideal mathematical mind). In order to justly assert a mathematical statement p one must have a (finite) construction procedure for p, in other words, one must have a proof of p. Thus, for the intuitionist, p is true iff it is proven. On the other hand, p is false iff it is incompatible with one of our constructions established earlier. That leaves room for statements which are neither true nor false. Goldbach's conjecture, stating that every even integer can be expressed as the sum of two primes, is such a statement. As neither has it a proof, nor has it a disproof it cannot be regarded either as true or as false. At least, not *at the current state of affairs*, it may not be truth-valueless eternally. (According to the so-called problem interpretation of Kolmogorov p is true if we have a solution of p [Kolmogorov 1932].)

It is clear that the principle of bivalence, suggesting that a well-formed statement is either true or false, cannot be maintained by the intuitionist. Moreover, it follows that the law of excluded middle ($\vdash p \vee \neg p$) also fails—just consider p to be the Goldbach's conjecture: nor the conjecture p nor its negation $\neg p$ can be asserted.[24] Generally speaking, the intuitionist abandons all the principles and methods he finds to be "non-constructive," that is, all the ones leading to general statements about our universe of discourse without the load of exhibiting concrete instances—thus such mascots like the law of excluded middle, the method of reductio ad absurdum, the axiom of choice, and so on. Nevertheless, on the positive side, he tries to "re-construct" as much of the universe (of mathematics) as possible with more rigor, more strict methods, less presuppositions.[25]

As no infinity can be built up by finite constructive methods, the intuitionist maintains quite different concepts of numbers, and an entirely different set theory. Individual natural numbers are taken for granted, but their totality is not: infinite sets are regarded as potential only. On the other hand, there is no problem with induction: each and every natural number can be reached by a finite, effective process. Thus each and every natural number is part of the constructive universe, even if there are no actual constructions of all of them, the demonstration of the possibility of those is enough.

It is by no way surprising that the intuitionist maintains a real number concept quite different from the classical one. An individual real number is often handled as a limit of a Cauchy sequence, a sequence $x_1, x_2, x_3 \ldots$, so that the elements the sequence get arbitrarily close to each other as the sequence proceeds: $|x_m - x_n| < \varepsilon$ (where ε is an arbitrarily small number). Clearly, a given real number cannot be identified with its generating sequence by the constructivist, simply because one cannot bear an infinite construction. Indeed, he regards these sequences as *infinitely proceeding sequences*, which cannot be exhibited in their totality, but one can always observe some initial, finite part of them.

Again, one can also assume that the first element of the other, not yet explored part of the sequence is given by *free choice*. It does not mean that the next element might be any rational number whatsoever; some restrictions may well be applied. For instance, one may restrict free choice to numbers satisfying the Cauchy property. This way at every point a dense host of potential free choice sequences is at hand, or as the intuitionist would put, a *spread* of them (of course, the classic set theoretical notions cannot be applied here). These choice sequences provide the base for the intuitionist concept of real numbers and the overlapping spreads for that of continuum. (I omit the technical details here.)

It follows that the intuitionist concept of equivalence also differs from the classical notion. To say the least, there are two ways for things not to be equal. If a and b are unequal, $a \neq b$, we can demonstrate a contradiction in supposing their equality. When a and b are *apart*, $a \#\!\!\neq b$, we do possess an integer k so that $|a - b| > \frac{1}{k}$.

A continuum, as conceived by the intuitionist, bears with some peculiar properties related to the classical one. It is clear that in a classical setup we can assert the following law (usually traded as the law of trichotomy):

$$\text{If } a, b \in \mathbb{R} \text{ then } (a < b) \vee (a = b) \vee (a > b).$$

In contrast, the intuitionist can make a weaker statement only:

$$\text{If } a, b, c \in \mathbb{R} \text{ and } (b < c) \text{ then } (a \#\!\!\neq b) \vee (a \#\!\!\neq c).$$

Probably the most striking feature of the intuitionist real number concept is that, because reals are exhibited as overlapping, ever becoming spreads, every function $\phi(x)$ from $\mathbb{R}$ to $\mathbb{R}$ defined on a continuous interval of reals is always continuous at every point. Thus, disturbing discontinuity examples, as the one described above, cannot be built on the constructive notion of real numbers.

Now, Dummett's offer is simply to regard those things as real, which can be accessed or constructed and regard properties real inasmuch they can be exhibited. We do not have to jettison real numbers from our scientific theories, all we have to do is to replace their classical concept with an intuitionist one. To make it short, Dummett calls for changing our account of reals, so to arrive to a less problematic description of the world which is more in harmony with our epistemiological ways. Nor would we have to miss analysis, a further good news, for Errett Bishop invented a constructive version of it (Bishop 1967). But the price to pay is not meager: we have to get rid of some of our deep, natural metaphysical convictions.[26]

This constructivist approach rhymes well with our concept of measurement procedures. When trying to exhibit some property of an object via measurement, we always arrive at some rational at a given point—or more exactly, to a result that can justly be represented by a rational number in any case. In a realist sentiment we will say that this result is just piece of data falling in what we really measure: an interval bounded by rationals, or our margins of error. Anyhow, we can always make our equipment more accurate and arrive at a number of another order, and so forth endlessly. This intuitive picture of our progress in accuracy is in line with the concept of free choice sequences: our steps toward precision are by no way lead by some necessity, our next limits in accuracy are entirely arbitrary.

In addition, we have now a way to lay down the theoretical burden of errors and margins. All we have is what we have access to. It is worth bearing in mind, when we ask nature these are her immediate and "unrepaired" answers.

Let us now get back to the thesis I began this chapter with: rationals are enough for measurement. After analyzing a thought experiment as a potential counter-argument, I concluded that for whatever difference (or system of differences) expressed by relative irrationals there is always a consistent renaming with rationals, preserving the relations arbitrarily beyond our exactness of our measuring equipment. When I talked about exactness, I relied on our realist intuitions: there is a real value somewhere out there which is approximated by our measurement operations. Now, after sketching the constructive scene, it is high time to get rid of these realist flavors and abandon the implicit commitment. Fortunately, it is not so hard to reword the conclusion without it. Say: I can always offer a renaming so that the differences of differences are well beyond the reach of our actual empirical (operational) means—whatever they might be. That is, I cannot *construct* any meaningful measurement for differences between the intervals picked out by irrationals and the intervals picked out by rationals as a result of my renaming. At least, not at the current state of art.

But so much for a general grounding of a constructive approach to measurement at this point. I will pursue the topic further toward the end of this book, in chapter 5. Now we will carry on with taking a glance on some further historical developments in measurement theories.

NOTES

1. Deciding which theory is streamlined enough is not a clear and straightforward task, as simplification at some point may result in more complex descriptions at others. Poincaré predicted that no scientist would ever choose non-Euclidean geometries for the model of space-time because of the resulting complexities (Poincaré 1902). As we know, he was not right in this respect. At this point, it can also be argued that we have no choice at all. The theories of Poincaré are identical (Szabó 2002, 26–37).

2. The practical implementations of these frequency standards are called fountain clocks (*Scientific American* 2004, Tal 2016). See some further details below.

3. Materials traditionally used for temperature measurement indeed show wide variety in thermal expansion (Chang 2004, 58, 109). Water, for instance, even expands when it is *cooled* from 4 to 1°C.

4. Triple point is a constellation of temperature and pressure where all the three phases of a material is present in equilibrium. An arbitrarily small change in the conditions would result in one of the phases of the material. The triple point of water is at 611.73 pascals and 0.01 °C.

5. In fact Campbell hoped to account for temperature as derived from genuine extensive quantities. Now with the expected inauguration of the new metric definitions in 2019 (CIPM 2017), his hopes may be fulfilled eventually.

6. It might not be entirely clear how to understand this, but here I omit the investigation, as it is not a major point in my argument.

7. Well, strictly speaking we may also need Campbell's error theory to "smooth out" data, but let us keep things simple.

8. Sometimes concatenation for temperature is conceived as a kind of averaging, which may be intuitive if one has the mixing of liquids in mind. However, mixing as a practical endeavor had never been successful as a base for a standard unit for temperature (Chang 2004, 60–68).

9. Indeed, sometimes we may have clear and cogent doubts about . Take the case of thermal expansion.

10. It is to be admitted that the whole empirical justification project is haunted by circularity. But "it is in the context of quantitative measurement, where . . . the problem of circularity emerges with utmost and unequivocal clarity" (Chang 2004, 221). (A bonus quote from Kyburg for the sake of interest: "In fact it is not clear what *is* error. There has to be some indication that the law of linear thermal expansion is reasonably close to being true before we can construct a thermometer as a way of measuring *temperature*" (Kyburg 1991, 85). (Emphases original.)

11. Observe that our theory encompasses the theory real numbers. Although Szabó would probably insist that we could tag with any kind of symbols we like, these symbols must align with an axiom system with high resemblance to the one for reals, not the least because we would like to tell stories on, e.g., acceleration in this framework.

(As it happens, he regards formal systems as genuine physical ones [Szabó 2003]. I addressed this view elsewhere [Csatári 2012].)

12. The name reflects that in these highly complex instruments particles are tossed up in vacuum tubes.

13. The idea of this thought experiment was put forward by András Jánossy in personal discussions.

14. We could, of course, also set up this thought experiment in a room with walls and ceiling, thus ensuring the one-to-one correspondence.

15. The chosen function preserves order, but it is not invariant under certain important operations. See more on this in the next chapter.

16. Roughly saying, by dense we mean that between any two values (however close to each other) there is a third value. The set of rationals is dense.

17. There may even be theoretical considerations pointing to this conclusion. Take *Planck time*, a natural unit stemming from an operational approach to space-time (Williams 2016).

18. Peter Forrest, one of the authors defending a discrete space-time thesis insists that there may be empirical means to support our choice (Forrest 1995). At least, in theory. His idea is based on a concept of systematic (i.e., theory-based) measurement and our ability to detect *systematic* errors in the data gathered by appropriately precise instruments (see chapter 5 for more on this). He admits, though, that the issue is open and may remain so for good.

19. The terminology used here reflects that of the papers of Michael Dummett (Dummett 1993 and Dummett 1995).

20. The empiricist may regard the statement in question *analytically true*, i.e., following from the mere logic and mathematics applied in the respective theory. But it only makes explicit that we do apply the theory of reals in our theory.

21. Though he does not explicitly mention Thomson (Thomson 1954), the essence of Dummett's modification is to bring continuous motion in the picture, while Thomson's lamp deals with two switch states.

22. To give a comprehensive account on Dummett's motives for a constructive model or language (discussed at length in his *The Logical Basis of Metaphysics* [Dummett 1991]) is far out of the scope of this text. For a concise summary see e.g., Kapsner's text (Kapsner 2014, 11–29).

23. For the purposes of this study, there is no need for a subtle distinction between intuitionism and mathematical contructivism in conceptual and historical terms, so here I use the two words as somewhat interchangeable.

24. For an intuitionist logical calculus, and also for an essential text on intuitionism, see Heyting's book (Heyting 1956).

25. For a concise account of the merits of constructive mathematics, see Bauer's paper (Bauer 2017).

26. It is worth noting, that Carnap also offers a choice between a language with classical features and one of a finitistic, constructive kind, as a matter of, so to say, free will (Carnap 1937).

Chapter 3

Scales and Structures

So far, while discussing measurement, we dealt with physical quantities. But what about social (or human, or special) sciences?[1] We may insist that having their own specific subjects, they must have their own specific quantitative concepts, if they ever want to measure. An alternative is opening toward the physicalist's way—and this could be a very popular direction nowadays had Papineau been right in that "we are all physicalists now" (Papineau 2009, 103)—with the admission that there are no non-physical processes. It is also often added that every concept of human sciences is reducible to ones of physics. And, having then the quantitative concepts of physics, why would we need new ones to measure? No doubt, one reason could be that such a reduction is hardly ever viable *actually*.

It may easily happen that the quantitative concepts of physics do not prove to be applicable in special sciences for whatever reason. Still, the vital need for keeping the precious tool of measurement remains in these disciplines. Thus, the practitioner of a special science is left with two choices. For one, she can verify that the received theoretical machinery of measurement is applicable for the given qualities of her discipline. That is, it is to be shown that they are quantitative concepts. Or, having failed with this, one can go for "liberalizing" the concept of measurement. But simply labeling things not labeled so before is not enough. It is much more apt to offer an alternative theory for measurement so to encompass new fields. That is exactly what the psychometrist S. S. Stevens offered. Among behaviorists "perhaps the most aggressive promoter of operationism" (Chang 2009) was Stevens. Social scientists and especially psychologists of the time found the spirit of Stevens' operationalism liberating[2] and it still has a major impact on these disciplines today.[3] It is somewhat understandable. For the behaviorists, building strict quantifiable concepts instead of the pre-existing sloppy ones accounted for by uncertain introspection was vital, and finding the proper philosophical grounds for this was a blessing.

To be sure, with Stevens, social scientists endorsed a version of operationalism more radical than that of Bridgman in many respects. Building his experimental work on auditory sensations on appropriate methodological grounds, he argued that discriminative, observable reactions to physical stimuli can be regarded as the measurable indicators of private experience. According to him, "to experience is, for the purpose of science, to react discriminatively" (Stevens 1935, 521). Interestingly, behaviorists' shipping away from individual experience was not after Bridgman's own heart. This fact may shed some light on why he "washed his hands of" Stevens (Chang 2009). Nevertheless, we can suspect that this was by no means his only reason.

As to his theory of measurement, Stevens' positivist-operationalist inclination is spectacular. First, he generously "liberated" Campbell's definition of measurement—measurement is the assignment of numerals to objects or events according to scientific laws—from all of its "fripperies." For him, any rule fits for the assignment, since, along with the logical positivists, he held that no mathematics is included in our observations, formal rules, as human inventions, constitute the framework of experience. Thus, our assignment conveys no natural structures, it is *structure generating* itself. It means that when we talk about the empirical operations of determining equality, more or less, equal differences and so on, as required by the establishment of the representational rules described above, we do not *determine* (exhibit) the specific relations between objects or events pre-existing out there, but we *determine* (stipulate) them ourselves (Michell 2007, 84–87). In other words, the specifics of a certain scale lie in the particular measurement operations we apply. This approach immediately cuts short the worries about the measurability of certain concepts of psychology. Concepts are defined by the very measurement operations. Hence the meaning of psychical qualities (i.e., their scientifically meaningful concept) lies in the way we measure them.

Surely, Stevens' work is justly criticized in many respects. But it undeniably opened (or let it, provoked) new vistas for measurement theories. In particular, his theory on measurement levels or scale types gave a rough nudge for the subsequent axiomatic foundation project.

3.1 PHYSICALISM AND THE PROSPECT OF MEASUREMENT IN SOCIAL SCIENCES

In his classical account of physicalism, Hempel takes psychology as a paradigmatic representative of the "sciences of mind and culture" (Hempel [1949

(1935)], 165). His main point is that every statement of psychology (unless it is a mere *pseudo-statement* lacking any content) has to have a translation into physical test sentences. These sentences contain only physical concepts, and are, of course, immediately ready for empirical verification. "The statements of psychology are consequently physicalistic statements. Psychology is an integral part of physics," just as every other empirical discipline (ibid., 168).

This philosophy, and it is not a typical fate for philosophies, found its massive support in the practice of science. It seems that behaviorism, as a movement inside psychology, is deeply engaged with the spirit of the physicalist view maintaining that the psychologist should turn her attention to the observable, bodily behavior instead of scouring about the secrets of mind through introspection.[4] All there is to be observed is physical behavior, and there is nothing to do with hypothetical concepts as mind, temper, or fear unless they can be reduced to observables.

But taking a closer look reveals some serious problems. According to Hempel's early, translationalist view just sketched, one can verify a statement of some special science by "translating" it first into physical sentences. Note that not only the actual verification is at stake here, since, according to the neo-positivist, verifiability is the very criterion of the *meaningfulness* of a sentence. (We may thus call Hempel's physicalism a semantic one.) The translation goes like this: "Paul has a toothache," a sentence involving a psychological concept, can be rewritten as different "test sentences," like, "Paul weeps and makes gestures of such and such kinds," "Closer examination reveals a decayed tooth with pulp," "Paul's blood pressure, digestive processes, the speed of his reactions, show such and such changes," "Such and such processes occur in Paul's central nervous system," or even, "At the question 'What is the matter?', Paul utters the words 'I have a toothache'" (ibid., 167). As Hempel himself observes, the list could be extended "considerably," which I would readily interpret as "without end." Hempel hastens to assure us that we do not have to verify all of these sentences, "some of them" is enough, and then we can infer the verification to others by induction.

That leaves us with a good deal of questions. How extensive should our always amendable list be? How many of the test sentences should be verified in order to say that we have verified the original statement? And to *what extent* should they be verified? Is it the same to take any of the sentences to verify? For instance, does the one on Paul's weeping count exactly as much as the one on Paul's nervous system?[5]

One may also wonder: do we justly talk about translation when trying to turn a sentence into an indefinite set of potential sentences? No wonder that later Hempel himself had given up the strict translationalist view and accepted a more liberal reductionist one instead, still maintaining that the concepts of special sciences must be reduced to physical ones. Everything that is psychical or social is realized in observable physical processes, even if the exact translation from the psychical-social realm to the physical one cannot be ensured. And at this point it is important to note that one can also be a physicalist while being anti-reductionist. That is, he can maintain that every *law* of social sciences is realized in physical processes, but deny at the same time, that these laws could be rewritten using only physical concepts (Fodor 1974).

I would like to make it clear here that the different directions of physicalism may have ontological, epistemological, semantic, or methodological flavors. Of course, usually these concerns overlap, but it is not without morals to see which feature is stressed in a given account. For instance, a physicalist may lay stress on one of the followings:

(a) Everything is physical (or supervenes on the physical).
(b) What can be known about nature can only be observed in physical processes.
(c) Every statement of the special sciences can be reduced to those of physics.
(d) (Therefore) every scientific method can be derived from the methods of physics.

Now it is interesting to see a major, historical debate on measurement as a debate among physicalists. The story also serves as a prequel to Stevens, introducing the context for his theory of measurement levels. The debate took place in the thirties in Britain. The British Association for the Advancement of Science appointed a committee in 1932, to "consider and report upon the possibility of Quantitative Estimates of Sensory Events" under the chairmanship of the physicist A. Ferguson. The members were noted physicists and psychologists of the time, Campbell was one of them. Joel Michell provides a remarkable account of the activity of the committee, the debate, its reports, and their context. I will rely on his texts in what follows (Michell 1999 and Michell 2007).

The starting scene of this debate was drawn by Campbell's account of measurement; in particular, as it was rethought and popularized Nagel and Cohen (Nagel 1931, Cohen and Nagel 1934). True, their stance was a bit more liberal than that of Campbell. Cohen and Nagel identify three uses of numbers: first, they can function as mere identification tags, second, they can mark a position on a degree scale, and third, they can represent quantitative relations of qualities (ibid., 294). The latter use, of course, characterizes what is traditionally called as fundamental and derived measurement. The second use, where one represents the relation of more or less by numbers, was also regarded as measurement by the authors. But they stressed that one must be aware of these different uses, since the different roles played by those numbers determine what can be *meaningfully* said about the measured qualities. For instance, as Russell put once, "to say that one degree corresponds to the same increase of temperature at any point of the scale, would be simply meaningless" (had he been right or not) (Michell 2007).

Likewise, it would be meaningless to say that a man with an IQ 150 is twice as intelligent than the one with a score 75 (Cohen and Nagel 1934, 298). All we can say that he stands higher in a specific performance scale. Worse, whether this scale represents a mere order structure is also a question of evidence. That is, it should be investigated whether intelligence as an attribute has an order structure at all.[6] It is at least questionable. And the situation has not been much rosier for other attributes of concern for psychology, but few arrived to the conclusion at the time that genuine measurement is not possible in the discipline.[7]

The physicists in the Ferguson committee had a strong opinion from the beginning to the end on the status of measurement: all there is to be measured are A and B magnitudes, that is, fundamental and derived ones—exactly as Campbell proposed. That means that alleged quantities like IQ or a sense intensity of a kind are not measurable, not even if they had been shown to exhibit orders. Reducing them to provenly measurable quantities would also be needed in order to talk about measurement.

By contrast, the psychologists of the committee insisted that this measurement concept is too narrow and it should be retailored so to fit the needs of special sciences. It seems that they ignored the fact that the measurement concept maintained by the physicists was based on *theory*. So instead of trying to work out an alternative, they rather regarded thc whole issue as a question of mere convention. To put it unfairly simply, they argued that the committee should accept a wider concept of measurement, because psychology measures, and measures differently than physics. Of course, we could even agree with the claim that concepts are conventional. But conventions are hard to change, and the Ferguson psychologists failed to convince the majority by appealing argumentation. However, I think they would have had a way to show that the concept maintained by the physicists is too narrow even by the standards of physics. There would have been an opportunity to address the problem of the measurement of intensive magnitudes that could not be reconstructed as derived measurement. In particular, they could have played around with the problems of measuring temperature.

According to Michell's account of the events, physicist themselves did raise the issue of temperature, and tried to present it as a B-magnitude (Michell 1999, 147–148). First, J. Guild made a not so fortunate observation that temperature is measurable "in a board sense" on the basis of some "arbitrarily postulated relations." But Campbell corrected his claim by saying measuring temperature is a genuine, simple derived measurement based on the Boyle-Mariotte law, which states that, in a closed system, the pressure of a gas is inversely proportional the volume if the amount and the temperature are held constant. Thus, the product of pressure and volume results a constant: $PV = NkT$ (where N stands for the number of molecules and T is temperature). Now, the order of this constant determines the order of temperature. But this constant encompasses to much. Temperature is thus not only dependent on pressure and volume, but the kind and amount of material as well. If we are not stuck to the celestial world of ideal gases, we must *discover* that the different values for this constant for different materials and amounts exhibit the favored structure. It is not easy, but the main point that it is to be shown. That could have been one card in the hand of the psychologists of the committee. Insisting that temperature measurement, vital to physics, is not convincingly proven to exhibit more than mere order would have been a good start. All would have been left is to show that sensation intensities did also exhibit orders. As we have seen, our empirically sound reason to maintain a temperature concept beyond mere order is based on congruent phenomenal behavior shown by the relentless non-theoretical experimental work of Regnault. Such operative-iterative ways are also open for social sciences, constituting a possible second step toward their quantitative concepts.

Michell shows that the psychologists in the committee missed another point, namely the problem of differences of differences (or second order of differences) in sensation intensities (ibid., 149–153). It is clear that once we are able to compare differences, we have a much more sophisticated structure than mere order, which bears the promise to be shown additive eventually. To put it concisely, psychologist failed to defend a suitable measurement concept, and physicists had easily won the battle.

Relying on my four physicalist theses above, I would give the following diagnosis. As implicit physicalists, the psychologists probably accepted (or would have accepted) (a) and surely (would have) accepted (b) of the above statements. But they probably failed to maintain the reducability of statements or concepts, (c), and therefore were reluctant to defend a strong opinion on methodological problems, to decide about (d). Michell even ventures that they exhibited a "curious indifference to methodological issues" (ibid.).

But social scientists' "reservations" about theorizing on science was hardly only a curiosity of the time. In a recent debate, Papineau admits: "the abstract metaphysics of physicalism may seem unlikely to have any concrete implications for the practitioners of the human sciences" (Papineau 2009, 122). Not in the least because even for the "hard" part of a special science, any reduction to physics would be so complex that it is hardly viable in practice. But in most cases, even the reducing theories are missing.

Even worse, special sciences may not only find certain methodological theories (of reduction, of measurement, etc.) futile, but also they often find them "inimical" (Shulman and Shapiro 2009, 127). They are instrumentalists, living on "middle-level generalizations," (Merton 1949) and do not like to see their disciplines constrained by inelastic theories. The best is to see reductive theories and other "games of the philosopher" as having nothing to do with the practice of special sciences. Science is living and let it live. As Shulman and Shapiro put it in a poetical note: "the idea that the certainty that accompanies theorems is the only hallmark of science is an obsolete hangover of the early Enlightenment" (Shulman and Shapiro 2009, 128). This is straight talk: we are really happy with the ways and tools of our playing around in our disciplines, please, do not even try to disturb.

Psychologists of the Ferguson committee had not yet had this self-confidence. Due to their physicalist inclinations, they were even hesitant to admit their tacit denial of methodological reduction. They just bitterly swallowed the conclusion of the committee: there is no sign of the measurability of sense intensities. Nevertheless, it was not the final word in the story: the stage was set for Stanley Smith Stevens.

3.2 INVARIANCE AND MEANINGFULNESS

To appropriately assess Stevens' contributions, it is worth taking a look on their ideological context. No doubt, logical positivism had a major impact on his views, and also, by no means independently of the former, operationalism (or operationism, in the psychologist's terminology). Stevens, as an advocate of logical positivism, often attended discussions and conferences featured by prominents of the Vienna Circle. He had high opinion about Carnap's book *The Logical Syntax of Language* (Carnap 1937); and he especially cherished the thought that mathematics as encompassing nothing but tautologies (i.e., analytic statements) cannot be caught in act in observations, but rather shapes the framework of experience (Michell 2007, 84). Building on this general background, in his work as a practicing scientist he relied heavily on the operationalist principles.

Stevens directly addressed and challenged the conclusions of Ferguson's committee and Campbell's measurement theory (Stevens 1946 and Stevens 1959). No surprise: their declarations explicitly denied the possibility of genuine measurement in psychology, and Stevens himself had worked out a scale for measuring the perception of sound intensity (the so-called sone scale). In his answer, he offers a new definition for measurement and as well a theory on the "levels" of the scales used in measurement.

According to a survey by Michell, many of the definitions of measurement circling around nowadays in the social scientific literature are closely related to the one of Stanley Smith Stevens from more than a half a century ago (Michell 2007). In Stevens' view measurement is assignment of numerals to objects or events according to rules (Stevens 1946, 677). No doubt, though the definition is paraphrasing Campbell's, this concept diverges from the empiricist one and seems to be rather "permissive."

First, it may sound rather strange, that measurement is applied to objects and events and not to their properties (or attributes). As Michell puts (a bit more sharply), it "makes no sense" to talk about measuring objects unless their attributes are measured (Michell 2007, 74).[8] Second, while Campbell requires the alignment with scientific laws, according to Stevens' wording, *any* rule fits. That sounds odd. We may be reluctant to esteem as measurement such a scenario, for instance, when one assigns the number six to the democratic countries, the number seventeen to the favorite pair of rubber boots of the Hungarian prime minister, and the primes to the factual mistakes of the commentators on the evening news in a respective order.[9]

Be the rules of the assignment however arbitrary, the applied rule affects the properties of the scale so gained, determines the "level" of measurement. Stevens specifies the following scale types: nominal, ordinal, interval, and ratio (see table 3.1) (Stevens 1946 and Stevens 1959). He shows for each level the respective empirical operation, mathematical structure, permissible statistical operations, and provides some examples. His main point is that there are different *structures* we can arrive at by measurement (as opposed to the Campbellian monotypic theory), thus numerals may have different meanings, among them ones seriously diverging from their usual understanding. We must always be aware of these meanings when dealing with them.

When creating nominal scales, one simply links a number to an object. We can arrive to such scales by establishing equality between certain objects, or by simply numbering the players on the football field. Note, however, that in this latter case it is not clear what the "equality" of player 7 in *a* team and player 7 in *b* team lies in. Along the same lines, my example above is a measurement by definition, though the common number of the prime minister's boots and one of the newscasters' mistake is not a product of our observing equality between them but of mere chance. No problem with it, may the operationalist reply, only we may find that these uses of numerals convey little information. According to Stevens, this kind of scale has scientific relevance in examples like giving numbers for occurrences of an event type or classes of objects because of some practical considerations. To bring an example from the practice of sociology, we use a nominal scale when assigning numbers to personal entries, such as codes for nationalities—even if I would be inclined rather to call it administrative coding than measurement.

Table 3.1 Levels of Measurement

Scale Type	*Operation*	*Structure*	*Statistics*	*Examples*
Nominal	Determination of equality	Unordered set structure (=)	Mode Number of cases	Numbering of football players Nationalities
Ordinal	Determination of greater or less	Ordered set (<)	Median Percentile	Street numbers Contentment
Interval	Determination of the equality of intervals or differences	Affine line	Mean Standard deviation	Temperature (Celsius) Birth dates Density
Ratio	Determination of the equalities of ratios	Field	Harmonic mean Logarithms	Length Temperature (Kelvin) Working hours for a Big Mac
Absolute	Determination of how many	Positive integers	Identity	Relative frequency

The mathematical structure of scales are to be characterized by the permitted transformations. In case of a nominal scale, this structure is given by $x' = \phi(x)$, where ϕ is one-to-one correspondence. This means that labels (which are numerals in this case) can be replaced, provided that things having the same label so far get identical labels, things having different labels before get different ones. The only statistic middle that can be applied to this kind of scale is mode; that is, we can pick the most frequent element.

Ordinal scales exhibit an $a \leq b$ relation on their values. The Mohs scale for the hardness of minerals, mentioned in section 1.2 (chapter 1), is often referred to as a typical example. According to this, m and n minerals are standing in $m \leq n$ relation, iff m can be scratched by n.[10] Many psychological measurements are said to belong here, like the ones dealing with the strength of sensations. Likewise, this kind of scale characterizes the grades in school and data sets as the one arrived at by asking interviewees to mark their contentment with the achievements of the government on a ten-degree scale.[11] The allowed transformation is $x' = \phi(x)$, where ϕ is any monotonic increasing function. Here, median can also be applied as a statistical tool, in other words, we are allowed to pick out the middle element from a finite data set, the one dividing it into two equal parts, upper and lower.

We talk about an interval scale when we have units, that is, when we can establish the equality of intervals or differences. This level is already considered as quantitative in an everyday approach as well. Temperature scales, Celsius or Fahrenheit, are usually mentioned here as examples, however, as we have seen, it is not at all trivial from the start that we can meaningfully talk about the equality of the intervals on these scales. Some say that the measurement of intelligence also exhibits an interval scale (or a near interval scale), but this sentiment is hard to defend.[12] I would rather offer the example of dates (measured from an arbitrary point in time), or direction in degrees on a compass. The characteristic mathematical transformation here is $x' = ax + b$, so that $a > 0$. The formula for the conversation between Celsius and Fahrenheit scales provides a good illustration: $F = 1{,}8C + 32$, where F stands for the given temperature on Fahrenheit and C on Celsius scale. At this level many statistical tools can be applied, but not those presupposing an absolute null.

The scale type just described is only one case for an interval scale, call it linear interval scale. But one can also specify so-called logarithmic interval scales, where the differences between certain ratios of attributes are quantified. Here, the allowed mathematical transformation is $x' = ax^b$, where $a > 0$ and $b > 0$. We could bring here the example of such derived magnitudes as density, or fuel-efficiency as dependent on the ratio of fuel-consumption (volume) and the covered distance (length).

The ratio scale is stronger than interval scale by the above mentioned absolute zero. Many of physical measurements are on this level, thus the measurement of length, mass, or even temperature on Kelvin scale (if we are convinced of the potential equivalence of differences). As for an example from the social sciences, measuring how many hours a worker may have to work to earn enough for a Big Mac in different countries could be mentioned here, though one may complain that the conceptual clarity of this example is far less than that of the ones from physics—not to mention the technical questions of measuring. Since the units of measurement are conventional, the allowed transformation is multiplication with a constant, $x' = ax$. At this level we already have the full range of statistical tools at hand.

It is common to supplement this list with the so-called absolute scale, where the unit of measurement is not arbitrarily chosen. Think of examples where pieces of something are measured eventually, such as in the case of measuring probability (as relative frequency), or the amount of Pirese[13] citizens in certain administrative datasets. On this level only the identity function is applicable as a mathematical transformation.[14]

One does not have to endorse Stevens' dubious definition of measurement in order to see that his theory of scale types is an important step in the right direction. Overlooking now the terminological question of what can be rightfully regarded as measurement and what not, an analysis of the structure of data acquired by our empirical procedures is of great importance. It is vital to see what is the conveyable meaning of the symbols used during measurement as assignment, and also to understand what is implied by the results. That is, our understanding of scales has immense methodological consequences, independently of whether the peculiarities of the scale types lie in our abilities (or inabilities) of setting up the proper procedures, or in the "real" nature of the observed qualities.

Stevens' taxonomy of scale types suggests the following considerations:

(a) Invariant properties clearly determine a given scale type.
(b) A given scale type clearly determines what can be meaningfully said based on the measurement data, what statistical tools can be legitimately used on it.

Let me begin with clarifying the first point. In the axiomatic tradition (discussed below in detail), by a scale we mean a function from a data structure to a numerical structure. Scales fall into different types according to their invariant properties. That means that invariant properties present necessary conditions for a type membership. Taking the above defined meaning of the scales with their intended properties (say, they have specific domains, namely empirically established structures), invariance is also sufficient for setting up a scale type.

By invariance we mean a system's staying constant in some of its properties under a given set of transformations. A square reserves its shape when mirrored by an axis running through one of its opposite angle pairs; the cardinality of an infinite set remains unchanged when a natural number is added; the patterns on a magic carpet stay the same as it's flying around over Baghdad; and the structure of data gained from some series of temperature measurement is the same be the values recorded in Fahrenheit or in Celsius. The figure is invariant under an appropriate mirroring, the size of the infinite set under finite addition, the carpet pattern under transport, the temperature scale under unit choice.

This latter observation is well in line with our intuitive understanding of measurement. May there be hot disputes on the conventional elements in our measurement procedures, no one questions that our unit choice is arbitrary. It is thus comforting to see that the outright conventional traits disappear in a definition of a scale type. Of course, not all of the scales involve units. So, it is interesting to see what invariance lies in, what are the preserved properties, and what are the eliminated contingencies for the elements of the above taxonomy of scales.

As mentioned above, nominal or classificatory scales are invariant under every one-to-one transformation. Here, to put it in a figurative way, class labels are preserved as essential to the scale type, but any constraints on the concrete titles on the labels are dropped, save that they must be different.

An ordinal or order scale is invariant up to every monotonic transformation. We do not even have to stipulate that these transformations are monotonic increasing, as Stevens did. Monotonic decreasing functions also do well. Consider a finite order with the assignment of natural numbers, and a function so that $f(0)=2, f(1)=3, f(2)=5, \ldots$, that is, the function assigns the nth prime to the nth element. But observe that $f(0)=-2, f(1)=-3, f(2)=-5, \ldots$, that is, ordering the negative primes respectively to the elements of the original scale does the job equally well, in the sense that, for example, $f(n)$ will always be between $f(n-1)$ and $f(n+1)$. In general:

$$f(x)\mathbf{R}f(y) \text{ iff } x\mathbf{R}'y\,.$$

It follows trivially that the transitivity, irreflexivity, and totality properties of the original order scale are preserved:

If $f(x)\mathbf{R}f(y)$ and $f(y)\mathbf{R}f(z)$ then $f(x)\mathbf{R}f(z)$;
If $f(x)\mathbf{R}f(y)$ then not $f(y)\mathbf{R}f(x)$;
For every x and y on the scale S, $f(x)\mathbf{R}f(y)$ or $f(y)\mathbf{R}f(x)$.

This reveals an important fact. Not only are the actual distances between the numbers used on an order scale arbitrary, but the "direction" of the scale, too.

A similar note is in order in regard to interval and ratio scales. According to Stevens, a characteristic transformation for an interval scale has the form

$f(x) = ax + b$, and the form $f(x) = ax$ for a ratio scale, since here we have a fixed zero point which cannot be modified. It is also assumed that $a > 0$. But we do not need this latter restriction, equal intervals are preserved even if we are "mirroring" these scales. However, we may wish to avoid zero length intervals, and insist that $a \neq 0$. But the moral is the same as before: in theory nothing prevents us from assigning lower and lower numbers to higher and higher temperatures; the "direction" of the scales is arbitrary. Despite all of these observations, we will stick to increasing functions in what follows, in line with the mainstream literature.

Finally, we have not yet accounted for measurement on the absolute level. Absolute scales are invariant up to identity, every other transformation "ruins" them. When we measure (count) how many votes are devoted for the different parties in a ballot box, our assignment of numbers is by no way arbitrary (though the numerals to represent them may well be).

As I indicated above, these transformations are not only characteristic in regard to a scale type, but also *definitive*. A scale type can be regarded as a class of certain individual finite scales arrived at by the appropriate measurement procedures. Now we can say with Patrick Suppes that scale types can be informally defined "as a class of measurement procedures having the same transformation properties" (Suppes 2002, 114–115).

Being so, invariant properties may also be suitable for identifying the given type of a given scale at hand. And it immediately leads us to our next challenge: meaningfulness. Let me reconstruct Suppes' example here (ibid., 110–111) with some modification. Suppose a psychologist who maintains that people's obtained scores on a test, $S(a)$, gives an ordinal ranking of their intellectual abilities, ia_a. Suppose further that the psychologist also records the age of each person tested, $A(a)$. Now, his hypothesis is that scale gained by dividing score with age, $iq_a = S(a) / A(a)$ also gives an ordinal (but age-compensated) measure of abilities, that is, whenever $ia_a\mathbf{R}ia_b$ then $iq_a\mathbf{R}iq_b$, and, of course if $iq_a\mathbf{R}iq_b$ then $f(iq_a)\mathbf{R}f(iq_b)$, for every monotonic transformation. Let us have the following data: $S(a) = 3$, $A(a) = 7$, $S(b) = 7$, $A(b) = 12$, $S(c) = 7$, $A(c) = 7$. Thus we have:

$$iq_a = \frac{3}{7}$$

$$iq_b = \frac{7}{12}$$

$$iq_c = \frac{7}{7};$$

that is: $iq_a\mathbf{R}iq_b$, $iq_b\mathbf{R}iq_c$, and $iq_a\mathbf{R}iq_c$. Now consider a monotonic increasing function φ, running on the denominators, which carries 3 to 6 and 7 to 8. This is a quite legitimate transformation, since the numbers as denominators represent data on the order scale. Now we have:

$$\varphi(iq_a) = \frac{6}{7}$$

$$\varphi(iq_b) = \frac{8}{12}$$

$$\varphi(iq_c) = \frac{8}{7};$$

that is: $\varphi(iq_b)\mathbf{R}\varphi(iq_a)$, $\varphi(iq_b)\mathbf{R}\varphi(iq_c)$, and $\varphi(iq_a)\mathbf{R}\varphi(iq_c)$. It is easy to see that the original structure is not preserved by φ, in other words, the truth value of the hypothesis is not invariant under this transformation.[15] But an empirical statement involving quantities can be regarded as meaningful only if its truth value is invariant under the appropriate transformations characterizing the respective scale type.[16]

Realizing the methodological significance of meaningfulness, Stevens clearly warns against the "illegitimate" use of statistical tools. As we have seen above, he not only utters a clear warning, but also tries to identify those tools which can be meaningfully used on a given level of measurement. For instance, when a social scientist measures some preference in a given population on a ten-degree scale, usually she has no reason whatsoever to presuppose that the differences between the pairs of neighboring degrees instantiate equal intervals. Therefore, it is not legitimate for her to conclude, say, that the interviewees like custard pie at a 5.36 points average—meaning the *mean* of the data by average. She can conclude, nevertheless, that they like it at 6 points average, meaning by this the *mode* of the data (6 was the most popular choice); or she can conclude that they like it at a 5 points, meaning by this the median of the data, or the middle point of the individual pieces of data ordered in a series.

It sounds nice. But unfortunately enough, when statisticians face a bunch of numerals, they are inclined to regard them as genuine numbers and they are ready to use all the methods generally used on numbers. They tend to be regrettably agnostic about the origins of their data, and thus the very meaning of the number-like symbols populating their database. Even when faced with this fact, they may shrug: "the numbers don't know where they came from" (Lord 1953).

Psychologists, social scientists, and statisticians likewise never liked the idea of statistical temperance. Despite all of these methodological minutes, the statistical armature produce *results*, they would say. We are not fools to tie our hands because all of these abstract theoretical reasoning (Velleman and Wilkinson 1993). This attitude may well sound familiar. It refaces again and again wherever critiques of prevailing scientific practices are presented on theoretical or methodological grounds. Spearman's theoretical considerations on mental abilities were received adversely by mental test practitioners (Gould 1996). Social scientists, though often uncertain in their methods, usually firmly reject any theoretical observations endangering their free exploitation of their usual ways (Shulman and Shapiro 2009).

Psychologists also brought complaints against Stevens' very typology: they said it was not exhaustive. They found that the theory of scales has an imperfect connection to the real world, for real measurement scales rarely exhibit the properties of a certain scale type perfectly—one often has to work with "intermediary" scales in some sense. In particular, some of them insisted that the most psychological data lay "somewhere between" order and interval scale. (Though, sadly enough, they usually failed to bring the evidence.)

Indeed, by a closer examination it becomes clear that the taxonomy is not exhausting. Consider, again, Mohs' scale of the hardness of minerals we met at the end of section 1.2 (chapter 1). It is not a hazy purple fiction at all to suppose a collection of minerals C with pairs of samples so that they do not scratch each other; that is:

There exist x and y in C so that not $x\mathbf{R}y$ and not $y\mathbf{R}x$.

Supposing, naturally, that there are also pairs in relation we have arrived to a partial order.[17]

This is a clear example for a scale missing from Stevens' list, but there are a good deal more. Louis Narens gives a formal account of scales and scale types, and tries to unearth the inherent causes why only certain scale types are favored by science despite the endless possibilities (Narens, 1981b). Also, our next section on the axiomatic approach reveals that the potential universe of measurement is stunningly immense, and only a tiny part of it is explored. But it will also be clear that purely a more comprehensive, more carefully built theory would hardly be the medicine to the social scientist's pains on measurement.

3.3 EXTENSIVE STRUCTURES

Axiomatic measurement theories have a long history. Their origins go back to Hölder or maybe even to Helmholtz (Hölder 1901, Helmholtz 1887). All the more surprising, the approach gathered real ground only after the watershed volumes of *Foundations of Measurement* (Krantz et al. 1971, Suppes et al. 1989, Luce et al. 1990) in the 1970s and 1980s.[18] Without a doubt, these results are of immense importance, some downright describe them as revolutionary (Cliff 1992, Michell 2007). Nevertheless, the high hopes "to inject these ideas into the mainstream of the behavioral and social sciences, just as was done with statistics" (Luce and Narens 1981, 214), cherished by the pioneering scholars—mostly behavioral and social scientist themselves—eventually failed to come true. The reasons are complex and will be explored in the upcoming sections in some detail.

Empiricist theories aim to provide the link between empirical procedures and abstract numerical structures through the rules of measurement. These theories fail, however, in providing any ready receipts for these procedures—not the least because they may differ considerably from case to case. Furthermore, the empiricist concept of measurement is uncomfortably narrow. By contrast, the axiomatic approach reveals a stunningly wide perspective for measurement, but, as a trade-off, it moves its focus from immediate empirical operations to the structures.

Students of axioms of measurement willingly endorse the claim that we assign numbers to the attributes of objects and events when measuring so to represent the relations of this attributes properly with the numerical properties. These relations can be exactly described by the language of mathematics, the measurement structures can be represented by axiomatic method. However, in reality, the order of things is reverse. First, axiom systems are worked out with an eye on the typical or not so typical measurement situations. Then the measurement theorist proves certain theorems, which reveal the relations of the possible models for a given axiomatic theory to familiar mathematical structures. Finally, it is up to the practicing scientist to relate her data to the available axiom systems, and, finding a match, to draw the allowed consequences. Nevertheless, fitting in data with structures is by far not a trivial task, as we may suspect and will see.

One of the first axiom systems for extensive measurement was given by the German mathematician, Otto Hölder (Hölder 1901). He provides a theory for an abstract, continuous, unbounded quantity, naturally instantiated by length. His concepts of magnitude and quantity is, unlike Russell's (Russell 1903), similar to the one adopted in this study. Hölder's system is characterized as follows.[19] For a given quantity Q, its magnitudes $a,b,c,d, \ldots$, and the classes of magnitudes A,B; and if for any three magnitudes a,b,c in Q, $a + b = c$ iff c is entirely composed of two discrete parts a and b, then the following axioms cover the system of extensive magnitudes:

(H1) For every a and b in Q one of the following is true:
  (i) $a = b$ (or equivalently $b = a$),
  (ii) $a > b$ (or equivalently $b < a$),

(iii) $b > a$ (or equivalently $a < b$).

(H2) For every a in Q there exists a b in Q so that $b < a$.

(H3) For every a and b in Q there exists a c in Q so that $a + b = c$.

(H4) For every a and b in Q, $a + b > a$ and $a + b > b$.

(H5) For every a and b in Q, if $a < b$, then there exist c and d in Q so that $a + c = b$ and $d + a = b$.

(H6) For every a, b and c in Q, $(a + b) + c = a + (b + c)$.

(H7) For every non-empty A and B classes of magnitudes of Q and every a in Q, so that

(i) for each a, a is in A and not in B or a is in B and not in A,

(ii) for every c in A and every d in B, $c < d$; there exists a magnitude x in Q so that for every magnitude x' in Q, if $x' < x$ then x' is in A, if $x' > x$ then x' is in B.

The intuitive meaning of these axioms can be given as follows. (H1) states that two magnitudes are either the same or different in that one is less than the other. (H2) ensures that there is no least magnitude. (H3) is that every magnitude is additive, so that the addition of them always results in a third magnitude. (H4) says that all the magnitudes are positive. (H5) assures us that that there is always a magnitude so to bridge the difference between a lesser and a larger magnitude; and there is always a magnitude with a difference to the larger magnitude so that exactly the lesser magnitude is needed to bridge. (H6) is that the addition of magnitudes is associative. Finally, (H7) ensures continuity by stating that no subclass of magnitudes with an upper bound has a least upper bound.[20]

Hölder proved that every magnitude in his system is measurable relative to any magnitude of the same continuous quantity, and eventually, that the realizations (i.e., models) of his axioms are isomorphic to the additive semi-group of reals. Inarguably, this is a strong result. Too strong, indeed. From a mathematical point of view it may describe an ideal structure for a magnitude, like a length, but not a structure for measurement operations. The theory can never be tested. Neither can we have a measuring rod arbitrarily small, nor can we have a set of rods so to always have a magnitude between any two magnitudes, however close. Not to mention satisfying the continuity property.

Again, as Patrick Suppes points out (Suppes 1951), Hölder's system has a property too demanding for a theory suitable for empirical test in itself, namely, it is *categorical*. It means that every two models of it are isomorphic; and, as we have seen, also isomorphic to a very strong structure. Suppes mentions another defect: Hölder treats '=' as the symbol of logical identity instead of using it for an equivalence relation properly axiomatized. Even when finding by measurement that the length of a whale is the same as the height of the national opera, we would be reluctant to regard the two properties as identical. And with good reason: they are properties of different objects distinct in space, we may be aware of the limits of "exactness" of our operations, "errors" in data, and so on.

These problems led Suppes to set up a new axiom system for extensive measurement (ibid.). It goes as follows. Let $\mathfrak{S} = \langle A, \preceq, \circ \rangle$ be a structure so that *A* is a set with the elements $a, b, c \ldots$, $\preceq$ is a partial order on the elements, and $\circ$ is an operation (which can be called concatenation). We can define the following by the $\preceq$ relation:

(i) $a \approx b$ iff $a \preceq b$ and $b \preceq a$,
and
(ii) $a \prec b$ iff $a \preceq b$ and $b \not\approx a$.

These bi-conditionals immediately make clear that equality is not treated as a logical constant, but defined by the only operation of the system. Now, the axioms of extensive measurement are the following:

(S1) If *a*,*b* and *c* are in *A* and $a \preceq b$ and $b \preceq c$ then $a \preceq c$.
(S2) If *a* and *b* are in *A* then $a \circ b$ is in *A*.
(S3) If *a*,*b* and *c* are in *A* then $(a \circ b) \circ c \approx a \circ (b \circ c)$.
(S4) If *a*,*b* and *c* are in *A* and $a \preceq b$ then $a \circ c \preceq c \circ b$.
(S5) If *a* and *b* are in *A* then not $a \circ b \preceq a$.
(S6) If *a*,*b* and *c* are in *A* and not $a \preceq b$ then there is a *c* so that $a \preceq b \circ c$ and $b \circ c \preceq a$.
(S7) If *a* and *b* are in *A* and $a \preceq b$ then there is a number *n* so that $b \preceq na$.

For the axiom (S7) we need to define what is meant by *na*. The definition goes like this:

(iii) $1a =_{df} a$ and $na =_{df} (n-1)a \circ a$.

Intuitively, (S1) claims for transitivity, (S2) guarantees that we always produces measurables by concatenation, (S3) asserts the associativity, (S4) the monotonicity, and (S5) the positivity of concatenation. (S6) stands for solvability, that is, it claims that there is always a concatenation for making unequals equal. Finally, (S7) along with its definition part guarantees that there is an arbitrary unit for the measurement, and stipulates the Archimedean property at the same time. The latter holds that there are no unmeasurably small magnitudes.

Suppes made important contributions in settling a couple of terminological issues. First, he introduced unique symbols for empirical relations and operations as clearly distinct from mathematical ones. Obviously, empirical and set theoretical structures are entirely different, and the very purpose of measurement theory is to show their possible relationships. Second, he gave some formal meanings to the terms *quantity* and *magnitude*. He called the above system the system of quantities. Magnitudes, according to him, are the equivalence classes defined by '$\approx$', which is thus a partition on A. He showed that while the system of quantities shows only homomorphy, the system of extensive magnitudes is "isomorphic to an additive semi-group of real numbers, closed under subtraction of smaller numbers from larger ones" (ibid., 169). He also proved that any such isomorphic semi-groups are connected through a similarity relation.

But Suppes was not reluctant to take notice of some weak points of his own system. His rather respectable conviction is that a good theory of measurement must reflect a structure recognized in flesh and blood empirical operations. Now, hardly any set of measures could be actually infinite in practice, as A certainly is. Moreover, general transitivity likewise cannot be enforced in real procedures, since, and here we might recall Bridgman's sentiments, all of our instruments are limited in sensitivity, precision, and scope.

The foundational project, initiated by Suppes himself among others, which resulted in the volumes of the *Foundations of Measurement*, embraces a general, comprehensive approach. The possible number of measurement systems is indefinite, still only three procedures of measurement lie behind them (Krantz et al. 1971). Generally they will be quite familiar, still I describe them here to indicate clearly the train of thoughts leading to an adequate axiomatization.

The first one can be called as ordinal measurement. When measuring ordinally, we apply some empirical procedure for matching certain attributes to each other, lengths of rods, for instance. Concatenation, that is, placing rods end to end by a straight line, has no role in this setup yet, all we require is a φ function assigning numbers the attributes so that $a \succ b$ iff $\varphi(a) > \varphi(b)$, where $\succ$ is the relation obtained by empirical matching, and $>$ is an ordering on the assigned numbers. The φ function guarantees, that if measuring c we find that $a \succ c \succ b$, then the number assigned to c will fall between the numbers assigned to a and b: $\varphi(a) > \varphi(c) > \varphi(b)$. In Stevens' terminology, with this procedure we can realize a measurement on an ordinal scale—well, if everything turns out right. For it can easily happen that we measure that $a \approx b$ and $b \approx c$, but $a \succ c$ (where $\approx$ denotes the relation that we did not ascertained difference while matching x and y), but this would imply according to the above lines that $\varphi(a) = \varphi(b)$ and $\varphi(b) = \varphi(c)$ but $\varphi(a) > \varphi(c)$ (where $=$ denotes the equality of the assigned numbers, of course), and this is clearly impossible.

Notice that here we are faced with the question of the infinite refinability of measurement. All right, we can offer a good advice for these cases: always take care of an accuracy for the measurement exceeding the order of differences between the pairs of attributes. But to be in line with this, we would have to obsess some preliminary knowledge about the order of magnitude of these differences, which we could gain by some measurement, in all conscience. We may assume further that (at least some of the) magnitudes are continuous, a fact triggering further questions—many of them were already addressed in earlier pages of this book.

The second procedure is solving inequalities—here concatenation enters the picture (which is again bad news in the face of our earlier investigations). It may happen that we have a relatively small set of data and find it impractical to apply standard sequences (to be explained in a minute). Then we can set up a couple of inequalities by matching the rods and their different concatenations against each other. By solving them we can acquire numerical approximations on the relative lengths of the rods.

For the third procedure type, let us make a quite strong assumption: there exist perfect copies (of rods in this case). Let us regard those rods as perfect copies of each other of which we cannot establish any difference. This assumption gives grounding for the procedure of counting units. If $a', a'', a''', \ldots$ are perfect copies, then we would like to assert the following: $\varphi(a' \circ a'') = 2\varphi(a)$, $\varphi(a' \circ a'' \circ a''') = 3\varphi(a)$ and so forth, where $\circ$ stands for the concatenation operation. The authors (and Campbell) call the $a, 2a = a \circ a', 3a = (2a) \circ a, 4a, 5a \ldots$ sequence a standard sequence. Thus we have a really strong tool: for any measurable b it can be asserted that for some n $(n+1)\,\varphi(a) > \varphi(b) > n\varphi(a)$. That is, be our standard sequence however chosen, there is always an interval between the unit's multiplication by an integer and its multiplication by the subsequent of the integer where b is found. In other words, any magnitude can be approached by an appropriately chosen unit. Moreover, as units are conventional and arbitrarily chosen, it also means that measurement can be infinitely precizified. (At least in theory. There may well be physical contingencies that pick out a shortest possible rod.)

The axiomatic approach aims to make explicit the assumptions we need on the empirical relation $\succ$ and on the empirical operation $\circ$ in order to be able to accomplish with the standard sequence procedure without contradictions, in other words, to construct the additive and order preserving φ homomorphism to the structure of reals. Thus, φ homomorphism is a mapping from the structure $\langle A, \succ, \circ \rangle$ to the structure $\langle R, >, + \rangle$, where R is a subset of the set of reals, $\mathbb{R}$. A *representation theorem* for a given axiom system states that there exists mapping from a given relational structure of measurement to a such and such numerical relational structure. Hölder's theorem states that the ratios of a continuous quantity is isomorphic to the additive semi-group of reals; the theorem proved by Suppes (usually honored also as Hölder's theorem) asserts the same for Archimedean property, which is weaker than continuity.

In addition there is another theorem to be proved from the axioms: *uniqueness theorem* in the *Foundations*' terminology. This is to deal with invariance; it gives the permitted transformation $\varphi \to \varphi'$, which characterizes the measurement procedure type (or if you like, the given level of measurement in Stevens' terms).

We should stop for a moment and reckon with the fact that while I dismissed the concept of concatenation as seriously flawed in the previous chapter, here we find it again alive and in good mood. The authors of *Foundations*, as their account on the procedures above suggests, indeed show some inclination toward a Campbellian sentiment. However, if we are ready to look over these paragraphs, a more liberal reading is possible. The exhibited structures never place constraints on the exact form of concatenation involved, nor do they regard it constitutive for the *meaning* of a given quantity. As I implied, there may be compelling non-circular ways of establishing units, and we might interpret concatenation here as an agreed way of unit manipulation so to arrive to an analyzable empirical structure, not as constitutive concept lying at the very heart of our measurement concept. Not in the least, as we will witness soon, the authors also describe structures with no direct reference on concatenation.

But let us now stay for a little while longer with the extensive structures. It is not without morals to see a refinement of the concept of extensiveness through formal analysis. The systems for extensive measurement were further advanced by several scholars, here I present a version of Luce and Narens (Luce and Narens 1981). Let $\mathfrak{X} = \langle A, \succeq, \circ \rangle$ be a structure, where A is a nonempty set, $\preceq$ is binary relation and $\circ$ a partial operation on A. $\mathfrak{X}$ is an extensive structure iff the following axioms hold for every a, b, c, and d in A:

(X1) $\succeq$ is a total ordering.
(X2) There exist a and b in A so that $a \succ b$.
(X3) If $a \circ b$ is defined, $a \succeq c$ and $b \succeq d$, then $c \circ d$ is defined.
(X4) If $a \circ c$ and $b \circ c$ are defined, then $a \succeq b$ iff $a \circ c \succeq b \circ c$; and if $c \circ a$ and $c \circ b$ are defined, then $a \succeq b$ iff $c \circ a \succeq c \circ b$.
(X5) If $a \succ b$ then there exists c so that $a \succ b \circ c$.
(X6) If $a \circ b$ is defined, then $a \circ b \succ a$ and $a \circ b \succ b$.
(X7) There exists a natural number n so that either na is not defined or $na \succeq b$, where nx is inductively defined as by $1a = a$, and if $(na) \circ a$ is defined then $(n+1)a = (na) \circ a$.
(X8) If $a \circ (b \circ c)$ and $(a \circ b) \circ c$ are defined, then $a \circ (b \circ c) = (a \circ b) \circ c$.

A definition is missing: $a \succ b$ iff $a \succeq b$ but not $b \succeq a$. Now, the first two axioms make sure that while the relation is total, there are different magnitudes. The third axiom is an interesting one, because it implicitly reveals that the operation is not necessarily defined for all of the element pairs of A. However, once it is defined for a pair, it is likewise defined for the pairs of the elements equal to or smaller than the original ones. (X4) claims for the monotonicity of concatenation. (X5) stands for restricted solvability. It is restricted, because instead of saying that there is always a magnitude by which unequals can be made equal, it says something like: every magnitude can be arbitrarily approached by the concatenation operation. (X6) is positivity: a magnitude gained by concatenation is always bigger than any of the elements in concatenation. (X7) brings the Archimedean condition and the definition of unit in one stroke. Finally, (X8) guarantees associativity.

In addition, if $\circ$ is a closed operation, that is:

(X9) whenever $a \circ b$ is defined, there is a c also in A so that $a \circ b \succeq c$ and $c \succeq a \circ b$,

then $\mathfrak{X}$ is a closed extensive structure. Such structures may be more in line with our intuitions on real measurement procedures: we are inclined to regard something gained by a concatenation of magnitudes a magnitude.

The other feature, by which this system is weaker than that of Suppes is more important: $\circ$ is not defined everywhere on $A \times A$, that is, not every pair of magnitudes is concatenable. This solution takes account of the operationalist concerns while preserving the conceptual unity of a quantity with what we might even call causal elegance. We cannot really measure either the diameter of a molecule or the diameter of the sun by concatenating our same old set of rods, still we can regard the quantity length as one and unique. Different measurement procedures define concatenation on different elements, still we are working with the same set of magnitudes.

A few more words are in place on representation and uniqueness theorems. I will omit here the lengthy and detailed proofs for (any of) the extensive system(s) described above, these proofs are provided for a broad family of measurement systems in the volumes of Foundations (Krantz et al. 1971, Suppes et al. 1989, and Luce et al. 1990). Below I only state these theorems and outline their proofs for a really simple system: finite weak order (Krantz et al. 1971, 14–17). This will do as an illustration how things work in the formal realm of measurement theories.

Weak orders are immanent in a wide range of measurement systems, though not in all. For a weak ordered system it is enough to suppose that we have a relation $\succeq$ on a set A for which transitivity and connectedness hold. (This implies that the relation is reflexive also.) These are really basic properties which are fundamental for physical measurement. But in measurements in social sciences they are often violated. For instance one can prefer b over a and c over b (e.g., because of quality considerations), still she may choose a over c (e.g., because of the too-large price difference). Preference is also not prone to be articulated for each and every pair of objects at issue.

We have two more requirements for our system to comply with: finiteness and non-triviality. It seems to be a natural and realistic presupposition that no dataset resulting from any measurement process could ever be infinite. So finiteness is really an integral part of any realistic measurement theory. However, infinite systems may be worth studying for theoretical interest. For one may suppose, say, that real systems' behavior tends to resemble to the ideal ones *in the long run*. As for non-triviality, we simply assume that our dataset is not empty.

Now, having our simple finite weak order system of measurement at hand we can state our two theorems.

(Th1) If A is a finite, non-empty set and $\succeq$ is a weak order on it, then there is a representation function φ (also called as measurement procedure, or scale) to a subset of reals with their usual order $\langle R, \geq \rangle$, so that $a \succeq b$ iff $\varphi(a) \geq \varphi(b)$.

(Th2) φ and φ' are both representation functions for the system $\langle A, \succeq \rangle$, iff there exists a strictly increasing function f so that $f(\varphi(a)) = \varphi'(b)$.

When considering the usual relation $\geq$ of reals, one can realize that it has (at least) another property beyond the ones listed for our weak empirical relation: it is anti-symmetric, so if $a \geq b$ and $b \geq a$ then $a = b$. A similar assumption is by far not trivial for empirical data: two records of the same value do not have to be regarded as identical. Fortunately, every weak order can naturally be linked to a total ordering. For this, we need to introduce two new relations by definition as the symmetric and asymmetric parts of $\succeq$ (just as we have seen above):

(i) $a \approx b$ iff $a \succeq b$ and $b \succeq a$.
(ii) $a \succ b$ iff $a \succeq b$ and $b \not\approx a$.

Now it is to be seen that the relation $\approx$ is an equivalence relation, that is, it is reflexive, symmetric, and transitive. As such, it determines equivalence classes: let us call **a** the equivalence class determined by a: $\mathbf{a} =_{df} \{b | b \approx a\}$. The so determined equivalence classes create a partition on the set A; split it into mutually disjoint subsets. The class of these equivalence classes is denoted as $A/^{\approx}$.

If it is the case that $a \approx b$, then, by the very nature of our representation function to be constructed, it follows that $\varphi(a) \geq \varphi(b)$ and $\varphi(b) \geq \varphi(a)$. Hence, by the antisymmetric property it follows that $\varphi(a) = \varphi(b)$. That is, every a and b in the same equivalence class must have the same scale value.

Now let us introduce a new relation $\succeq_p$ on the elements of $A/^{\approx}$. What is left is to construct a function ψ so that $\mathbf{a} \succeq_p \mathbf{b}$ iff $\psi(\mathbf{a}) \geq \psi(\mathbf{b})$. Here, as our partition is countable, and what is more, finite, we can create our function simply by counting. So let us choose $\psi(\mathbf{a})$ so to pick the number of **b**s (where **a, b** $\in A/^{\approx}$) so that $\mathbf{a} \succeq_p \mathbf{b}$. If it is the case that $\mathbf{a} \succeq_p \mathbf{b}$, and there is a **c** so that $\mathbf{b} \succeq_p \mathbf{c}$, than also $\mathbf{a} \succeq_p \mathbf{c}$ by transitivity, that is, **c** is counted for both **b** and **a**. It, then, follows that $\psi(\mathbf{a}) \geq \psi(\mathbf{b})$. On the other hand, if not $\mathbf{a} \succeq_p \mathbf{b}$, then $\mathbf{b} \succeq_p \mathbf{a}$ and $b \succ a$; so there is a **c** counted for **b** but not **a**. Hence $\psi(\mathbf{b}) > \psi(\mathbf{a})$. By this, (Th1) is proved.

Our uniqueness theorem—which would be more apt to be called invariance theorem—states that our system is invariant up to every strictly increasing function. It is easy to see that if it is the case that $\psi(\mathbf{a}) \geq \psi(\mathbf{b})$ it is also true that $\psi'(\mathbf{a}) \geq \psi'(\mathbf{b})$, if $\psi'(\mathbf{a}) = f(\psi(\mathbf{a}))$, where f is any strictly monotonic, increasing function. If it is not the case that $\psi(\mathbf{a}) \geq \psi(\mathbf{b})$, then $\psi(\mathbf{b}) \geq \psi(\mathbf{a})$, and then also $\psi'(\mathbf{b}) \geq \psi'(\mathbf{a})$ by the pattern above. We can observe that transitivity is preserved, and it is somewhat trivial to realize without going into formal details that no different values will be "concurred" by f, since the strict property guarantees that all distinct elements in the domain will have different images. What is left to show is that no weaker constraints on f are enough to suffice as a uniqueness function.

We can weaken the constraints on f by presupposing that it is monotonic, though, but not strict. In this case different $\psi(\mathbf{a})$ and $\psi(\mathbf{b})$ on the domain may have the same image: $f(\psi(\mathbf{a})) = f(\psi(\mathbf{b}))$. It then follows that $f(\psi(\mathbf{a})) \geq f(\psi(\mathbf{b}))$ and $f(\psi(\mathbf{b})) \geq f(\psi(\mathbf{a}))$. But the latter should not be the case. We supposed that $a \succ b$, so a function with the image $\psi'(\mathbf{a}) = \psi'(\mathbf{b})$ is unacceptable. Hence our rough proof for (Th2) is concluded.

Of course, weak and total orders are among the most simple structures. Proofs for the respective representation and invariance theorems can be much more sophisticated for finer systems. And, as it happens, we have quite a lot of them.

3.4 VARIATIONS ON AXIOMS AND REPRESENTATION

A wide scale of structures of extensive or partial extensive features has been studied over time. Changing the properties or even the number of operations and relations involved may reveal new systems of empirical interest, but could provide new insights even into the previously well-known structures.

Thus, it can be shown that the generally supposed condition of total ordering can be weakened to a reflexive transitive relation, and the representation for additive reals can still be shown under some circumstances. What is more, even an equivalence relation will do if the Archimedean axiom is strengthened (Luce and Narens 1981).

It is also of interest to see what happens when the Archimedean axiom is abandoned. It is a *necessary* axiom, meaning that systems lacking it have no additive representation on the reals. However, they have on different mathematical structures, for example, on non-standard reals. One may wonder how justly these representations could be called numerical, and hence, whether there is a natural class of representations to be called as measurement. I will address this point below.

An important class of theories is constituted by the systems where the conditions on the concatenation operation is weakened. Abandoning (X8), associativity, we arrive at the class of positive concatenation structures. It is shown that these systems exhibit really strong invariance properties: whenever φ and ψ are representations and have the same range, and there is a function f so that $\psi = f(\varphi)$ has a fixed point, then $\varphi = \psi$.

Weakening the properties of the operation further, we arrive at the intensive structures. Here the main point is dropping (X6), positivity. (X7), the Archimedean axiom must also be replaced by a variant. For some important intensive structure (consider, e.g., the measurement of temperature), the positivity axiom can be replaced by the following:

(Int) if $a \succ b$ then $a \succ a \circ b \succ b$.

Empirical operations like this can be conceived intuitively as a kind of averaging.[21]

It remains to be seen that measurement data are not necessarily one dimensional. If this is the case, the underlying set is a Cartesian product of two or more sets. It is natural to maintain that the relations and operation belonging to different dimensions are different in nature. This train of thought leads to a realm of structures of wide variety, many of them with useful morals especially for measurement in behavior and social sciences. Below a relatively simple system is considered.

Let $\mathfrak{C} = \langle A \times P, \succsim, xy \rangle$ be a structure. It is a conjoint structure solvable with respect to the element xy iff xy is in $A \times P$ and the following axioms hold:

(C1) $\succsim$ is a weak ordering on $A \times P$.
(C2) There exists ap in $A \times P$ so that $ap \succ xy$.
(C3) For each ap and bp in $A \times P$, if $ap \succ bp$ then there is a c in A so that $ap \succ cp \succ bp$.

(C4) For each ap in $A \times P$ there exist b and q so that $ap \approx bq$ and $ay \approx xq$.

(C5) For each a and b in A so that $ay \succ xy$ there exists a positive integer n so that $(na)y \succ by$; where na is defined inductively as follows: $1a = a$ and na is defined and s is such that $ay \approx xs$ then $(n + 1)a$ is some u so that $uy \approx (na)s$.

(C6) For each a and b in A and p an q in P;
 (i) if $as \succeq bs$ for some s, then $ap \succeq bp$,
 (ii) if $wp \succeq wq$ for some w, then $ap \succeq aq$.

(C7) $\succeq_A$ and $\succeq_P$ are total ordering on A and P respectively, where
 (i) $a \succeq_A b$ iff for some s, $as \succeq bs$;
 (ii) $p \succeq_P q$ iff for some w, $wp \succeq wq$.

An intuitive reading of the axioms can be given as follows. (C1) is, of course, weak ordering as it rightfully asserts of itself. (C2) is non-triviality, there are different elements in $A \times P$, which are not equivalent. (C3) states that the asymmetric part of the relation is dense. (C4) asserts solvability with respect to xy and (C5) is the Archimedean axiom. (C6) is independence, stating that a relation between two elements with one common component remains however we choose that component. Restricting this way the relation to one or the other component, (C7) guarantees that these derived relations are total orders on the respective component sets.[22]

What conjoint structures are supposed to show is that we can measure qualities or properties indirectly by the assessment of two (or more) component the wanted quality depends on. But up to this point, it is no stunning news. Of course, we can measure velocity by measuring its (definitive) components: time and distance. The point is that in these systems we can verify that the required property or the components are quantities in the sense that they bear a meaningful unit concept, simply by certain relational constellations, without relying on the concept of concatenation. Again, conjoint systems show that *quantifying certain non-physical qualities is logically possible*. It is easy to interpret them to suggest that *whether a quality can be quantified is an empirical question.*

Practically, this certainly means that the axioms must be verified on the data. The independence axiom, (C6), also called as single cancellation is one of crucial importance. Suppose that we are to measure thermal comfort, supposing it to be dependent on the two components: heat and humidity.[23] Now if we found that higher temperature is always accompanied by higher comfort however we choose humidity, and also, lower humidity is always accompanied by higher comfort wherever we fix temperature, then we could conclude that single cancellation axiom is satisfied by the measurement data.[24] If so, our system fulfilled a necessary condition for calling it measurement (beyond mere order).

But this condition is not enough, axiom (C7) must also be fulfilled. A version of it, more apt for empirical tests on real data, is called double cancellation axiom. It provides a means to identify relations across different levels of the components, and would eventually require—not so surprisingly—for the components to turn out to be quantities themselves. Only this could guarantee the quantification of the conjoint quality in question. However, coping with this axiom is still not sufficient, all the other axioms are needed with special concern on (C4), the solvability and (C5) the Archimedean criterion.[25]

Here we have an immediate moral for our present investigations. In my terminology, the method of conjoint measurement exhibits phenomenal congruence in a body of data. It follows that Chang's epistemic iteration is not the only valid way to operationally introduce a unit. The concept of phenomenal congruence goes beyond the primarily perceived events and their instrumental manipulation. Having realized this, one might suppose that the axiomatic method should have had a global fertilizing effect on scientific practice. This was not the case, and some of the reasons are explored below. But before that, some more words on axioms and justification are in place.

3.5 THEORY AND TESTING

Presenting scientific theories as axiom systems is a widespread practice nowadays. Formal treatment helps make clear what is and what is not entailed by a theory, and thus, not the least, it provides more room for empirical confirmation (or falsification). At the same time, it offers no remedy for the well-known problems of justification, some of which will be briefly sketched in this section.

After Carnap, Szabó describes a theory as an (*L*, *S*) pair, where *L* is a formal system and *S* is a semantics, tying the consequences (theorems) of the system to the empirical facts of the world—also called as correspondence rules in the pristine terminology (Szabó 2013). Let us first examine how a formal system is built up. *L* consists of a language, some rules of derivation and the axioms. Axioms come in different groups, at least as long as their origin is concerned. Let *L* be a part of a physical theory. The first group of axioms is formed by those of logic. Usually classical first order logic is concerned, and these axioms are often not made explicit.[26] A second group of axioms encompasses the mathematical ones. This group may vary depending on the mathematics applied in the given theory. Only after this can we construct the physical axioms, which (together with the previous ones) grab (hopefully) every intended aspects of the given physical theory. But finding the appropriate axioms, especially the "empirical" ones, is not a trivial task at all. No strict method exists, we can mostly rely on intuition and the "method" of trial and error.

Szabó (2013) denies the essential differences between the axiom groups above. Differences in their heuristics is also not a really convincing reason for the grouping: the underlying logic and mathematics may well be revised while shaping a theory. All in all, one may insist that sorting the axioms like above is entirely arbitrary. Certainly, all of the axioms could be merged into one by concatenating them with conjunction. So the number and style of axioms is matter of taste, and serves only conceptual clarity. For instance, as we have seen, (X1) simply calls for a total order in words, but, of course, can easily be split into more elementary, more formal axioms, as total ordering requires the transitivity, antisymmetry, and totality properties ((X1a), (X1b), (X1c) respectively):

(X1a) If $a \succeq b$ and $b \succeq c$ then $a \succeq c$.
(X1b) If $a \succeq b$ and $b \succeq a$ then $a \approx c$.
(X1c) For every a and b, $a \succeq b$ or $a \succeq b$.

Surely, less trivial examples could be drawn by making the hidden axioms of logic explicit and combining them with the "high level" axioms. But anyhow, even if Szabó is right about the equal epistemiological status of the axioms (of which I am not convinced [Csatári 2012]), a properly structured axiom system is inevitable for transparency and understanding.

I would like to stress here that not all of our underlying presuppositions and inclinations are necessarily tracked even after the most elaborated axiomatic utterance. Take Chang's example of an ubiquitous ontological commitment in context of measurement, the *principle of single value*: "a real physical property can have no more than one definite value in a given situation" (Chang 2004, 90). Neither it can be squeezed from an axiom system for measurement, nor it is stated as theorem. Interestingly, this metaphysical principle is still present in the system: the very notion of function takes care of it.

While many of the axioms are products of tiresome scrutiny after the intended strains of a theory, some of them are readily given. As we are in any case after a representations theorem when talking about measurement, a homomorphism into a subset of reals, some of the axioms are constrained by the very nature of reals. Axioms (X1a), (X1b), and (X1c) are all necessary in this sense. Natural ordering on reals has these properties. (Reflexivity is also a necessary axiom, but it is implied by the total order, so there is no need to state it separately.)

By contrast, other axioms are not necessary in this sense, but serve to exclude trivial cases, limit the extent of the system, or guarantee a fine-grained structure for the empirical theory. Thus (X2) ensures that there are at least two elements in relation. Again, the requirement of finiteness is a quite strict bounding of a given theory, but clearly a typical feature of measurement data. Finiteness is not always stated as an axiom, it may also be given in the respective Hölder-style theorems, but it is just a question of taste.

Existential axioms, like the solvability axiom (X5), provide example for structural non-necessary axioms. Clearly, nothing guarantees that for any two elements in an arbitrary subset of reals we will always have a third element exactly bridging the difference between them. However, it is true for some "natural" subsets such as integers.

In most of the cases Archimedean property (X7) is also presupposed. It is clear that this property holds for the additive field of reals, thus this axiom is necessary for the systems homomorphic to it. What is strange that it needs to be stated, it seems to be independent of the other axioms (Krantz et al. 1971, 25–26). All the more sad, this axiom has some quite unpleasant properties. To see this, notice that the property can be reworded in the realm of measurement structures as follows: *every strictly bounded standard sequence is finite*. It is trivially true for finite structures, that is why it is a bit surprising the axiom is needed. On the other hand, there is no way to *falsify* it for infinite structures, because no one ever generated an infinite standard sequence.

At this point I feel a need to expose briefly some of the notorious philosophical issues that concern justification. Nothing new is intended to be said at this point, I only want to foster a deeper grasp of the problems of theory and testing.

Rudolf Carnap made an important conceptual distinction by differentiating between statistical and logical probability (Carnap 1945, Carnap 1966, 19–39). By the first he meant the so-called *frequentist* interpretation worked out by von Mises and Reichenbach (Mises 1939, Reichenbach 1949 [1935]). To this view probability is nothing else than the relative frequency of the favorable events in a given event series; or rather, the limit of this frequency in the long run. This approach gets rid of the vicious circle intrinsic in the so-called classical approach. According to the latter, the probability of a favored event is determined by the number of the *equipossible* cases. But the obvious interpretation of equipossible is equally probable, which is circular. Gamblers might not welcome the frequentist view as they needed prompt estimations instead of time-consuming experiments, but Carnap found this concept quite apt for scientific purposes, and indispensable in the description of phenomena of statistical nature.

Entirely distinct from this, based on the works of Keynes and Jeffreys (Keynes 1921, Jeffreys 1939), Carnap argues for a different but equally legitimate notion: logical probability. This latter accounts for the *degree of confirmation*. So while statistical probability is an empirical concept describing states of affairs in the world, logical probability describes the relation between linguistic entities: a hypothesis and a confirming (or disconfirming) statement of evidence.

Carnap began to develop a formal system for handling this kind of probability in his 1950, and worked out a solution for simple languages (Carnap 1950). New puzzles and paradoxes of induction, however, like Goodman's or Hempel's fatally weakened the position of his approach. Let me briefly summarize these latter two.

Goodman's famous riddle is the following (Goodman 1955, 72–81). Let us call something "grue" if it is examined before a certain time *t* and is green or not examined until a certain time *t* and is blue. Now how would empirical data on some green emeralds observed before *t* confirm the following hypotheses?

Every emerald is green.
Every emerald is grue.

Goodman identifies projectible predicates and non-projectible ones and also law-like statements and non-law-like statements. The main problem is how to distinguish them. Anyhow, what can be seen for sure is that the confirmation of a hypothesis "depends heavily upon features of the hypothesis other than its syntactical form" (ibid., 72–73).

Hempel's paradox (or the raven paradox) was first worded by the Polish mathematician Janina Hosiasson-Lindenbaum (Hosiasson-Lindenbaum 1940).[27] The paradox goes like this. Consider any law-like generalization, for instance:

All ravens are black.

Formally:

$$\forall x(Rx \supset Bx)\,.$$

Now, this hypothesis is confirmed by each black raven. The problem is, that the above expression is logically equivalent with the following:

Everything that is not black is not a raven.

$$\forall x(\neg Bx \supset \neg Rx).$$

But this is confirmed by all non-black non-ravens, for example, by all snowmen—a result, though not really a paradox, stunningly counterintuitive.

In the wake of these disheartening results it became clear that inductive logic cannot be formalized the way deductive logic can be, simply because the validity of the inductive inference cannot be judged by its mere form. All in all, confirmation can only based on the symbiosis of our ever changing, ever interacting "practices and standards."[28]

While trying to confirm our axiom systems, regarding individual axioms as hypotheses, we must face the fact that we do not have strict, formal, let alone automatic methods at hand, we must rely on our experiences, common sense, and practical considerations. In the particular case of confirmation of axiom systems as theories of measurement, many of the difficulties are related to the fact that existing datasets are always imperfect in the sense that they lack some of the ideal models' features. To begin with, they are always finite, whereas axioms systems usually have infinite models.[29] Of course, we can guarantee finiteness by the axioms. But even having finite systems, some of their interesting properties may hardly be confirmed.

Consider measurement on cyclic scales, such as time measurement on a twelve-grade dial. Suppose naturally that our units can always be divided into subunits. As it happens, it will be derivable from our axioms that for our any two readings, however close to each other, there is a (potential) reading between them. It is definitely not against our intuition about the progress of the clock hands. This property, though, can never be *satisfactorily* confirmed, be our dataset of actual readings however rich.

A major problem is that while certain axioms are quite easy to confirm or disconfirm, others may well resist to trying (Krantz et al. 1971, 28–30). Take an axiom granting transitivity, for the sake of simplicity (S1):

(S1) If a,b, and c are in A and $a \preceq b$ and $b \preceq c$ then $a \preceq c$.

Now suppose we have a rich record of measurements on some relative property of a bunch of objects, among them x,y, and z. According to the records, we found that y is, say, *brighter* than x, and z is brighter than y, still, x is brighter than z. Suppose we found similar results for other triples also. In this case we can conclude that a measurement theory involving transitivity is (to a great extent, to a high probability, etc.) disconfirmed by the available data.

Now take the Archimedean axiom:

(S7) If a and b are in A and $a \preceq b$ then there is a number n so that $b \preceq na$.

Disconfirming this axiom is much more problematic. For not finding the proper n for a given pair does not mean that such an n cannot be exhibited at all by the measurement procedure instantiated by the data.

One might say that finding no refuting objects provides a kind of confirmation for the hypothesis (and thus for the theory),[30] and many are inclined to follow this principle in practice, indeed. Not to mention that we may not be better off with "positive" confirmation either. Take, for example, the axiom for solvability:

(S6) If a,b, and c are in A and not $a \preceq b$ then there is a c so that $a \preceq b \circ c$ and $b \circ c \preceq a$.

It is clear, that confirming this axiom can be hard even on not too large datasets. First, simply there are too many cases which cannot be checked within reasonable time. Second, even finding unresolvable cases does not automatically mean a refutation for the axiom, we may simply blame or "chosen" data (set of objects) for a given measurement procedure. To put it in another way, a set of data may be large enough to have a check on it for a property in reasonable time. On the other hand, a set of data is almost always too small to include a "solution" for every pairs, for instance.

In practice, as Krantz et al. (ibid., 30–31) note, scientists—those who happen to dive into such activities at all—tend to be generous while verifying axioms. Non-necessary axioms, for instance, are rarely tested. By necessary axioms, as explained above, we mean those simply entailed by a homomorphism to the required mathematical structure—most commonly to some substructure of reals. Non-necessary axioms account for more subtle, contingent features as finiteness, or even solvability. No doubt, the problem of the confirmation of the solvability axiom is easily circumvented by not testing it at all.

It may even happen that while all of the axioms (hypotheses) are confirmed (to an acceptable degree), there is still a *consequence* of them which fails to agree with the data—a further knotty problem. How could it happen? The thing is that all of our confirmations are partial, at best they can be asserted with a *rather high* probability. But we can never reach the certainty of deductive inference. What is a consequence of an axiom system is a deductive consequence of it and it can easily be imagined that imperfect data violates some inferred feature on the long run.

Despite this, it could be a good feedback to test some consequences of axiom systems. But how to choose from the infinite many implications? Sure, complexity is a living criterion: we will choose from the simplest ones and not from the ones of immense complexity. For which we have no reason whatever beside mere practical ones.

All that is too familiar, one might say. It is well known that perfect verification is never viable, we must do with confirmations of different degree, but here we are just facing with the same old questions: how do we know that an axiom is "confirmed enough?" Let alone: how do we know *to what extent* is an axiom confirmed? Neither in numerical terms, nor for sure can we tell this. All we can rely on is our practical ways aided by some sober considerations on theories and instruments. But that does not mean that the task of verification can be taken lightly; on the contrary, it needs tiresome and careful work to give our best guesses.[31]

NOTES

1. In general, in this chapter I will talk about social sciences in a broad sense. In particular, I will include psychology by default. A debatable choice though, it makes sense as long as similarities exceed peculiarities. I will, of course, be specific whenever needed.
2. Indeed, the word "operationalism" (or "operationism" in its original form) itself was coined by the renowned experimental psychologist, Edwin Boring (Chang 2009).
3. There is no consensus on the assessment of this fact in the contemporary literature. For instance Bickhard (Bickhard 2001) (in his reflections on Grace [Grace 2001]) regards it as a curse for the methodology of psychology, while Feest (2005) argues that the operationism of the practicing scientist is misinterpreted as a theory of meaning or knowledge, thus the arguments that it fails as such and such a theory miss the mark (Feest 2005).
4. For one with an apt historical rigor, it may sound anachronistic to use the word "physicalism" in the context of behaviorism, which has its roots in the late nineteenth or early twentieth century, while physicalism as a philosophical commitment origins in the work of some members of the Vienna Circle in the thirties. By the same token, it is likewise anachronistic to use the word in context with Campbell. I will be consequently anachronistic in this respect.
5. Note that this is just mirroring a general problem of verification: is it equally reasonable to verify any prediction of a theory? Surely, we well may have arguments that not.
6. However, one may insist that intelligence is nothing more than the ability of dealing with IQ tests. Then, the least we can say that this ability is higher for one who scores more on the test and lower for the one scoring less. (More to be said on the operationalist approach below.)
7. Michell mentions the psychologist H. M. Johnson as one exception, and also notes that his work was widely ignored. (Michell 1999, 142 and Michell 2007, 82)
8. On a more abstract level, it is not nonsense to insist that measurement is "about" objects. For instance, Kyburg defends this view. He suggests to conceive quantities as functions such that their domains are sets of things and their ranges are the real numbers (Kyburg 1984, 17).
9. Michell mentions two more problems with this definition, but both of the issues are already present in Campbell's concept (Michell 2007, 74–75). For one, according to Michell, measurement is concerned with *numbers* as relations or ratios between the magnitudes of attributes, *numerals* are needed only as words, they are not essential.

True, it is not self-evident to establish ratios among spoken number words, tin plate street numbers, playing cards, and digits on seven-segment scoreboard displays. But, of course, we regularly use numerals while measuring, and there is no problem with it, as long as we use them *systematically*.

What is more, there is also a reason to defend Campbell's numerals as subjects of assignment beyond his own considerations on the status of numbers. We can regard numerals as arbitrary symbols by which we represent some datasets compiled as results of our measurement procedures. Now we can never take as granted that these datasets or even their idealized infinite expansions will bear the same structure as, say, the field of rationals. That is, numerals have different meaning when we assign them for measurables, than when we use them as symbols for mathematical objects. This is one of the main morals of this and the upcoming sections.

Michell's other point is that measurement is about discovery and knowledge, and cannot be simply regarded as the assignment of symbols. But generally, nothing prevents us from trying to convey this knowledge by the act of assignment.

10. The example would be really apt only if we could guarantee the transitivity and connectedness of the scale, which we cannot.

11. Though, as we will see, further problems arise with the preference or sensation scales.

12. For a vivid and rather controversialist history of IQ measurement, see Gould's book (Gould 1996).

13. A fictional nationality made up for surveys on xenophobia in Hungary.

14. It is worth a mention that Stevens' system was extended or modified several times by several authors over time. For instance, Mosteller and Tukey offer an improved system of levels with better fit for some scales used in scientific practice, such as probability or percentage (Mosteller and Tukey 1977). Chrisman presents a taxonomy more in line with the practice of cartography. As these systems carry little additional philosophical interest, I do not address them in this text (Chrisman 1998).

15. Suppes' example is snappy because it reflects a highly debated issue of psychometrics with long history. Indeed, $S(a) / A(a)$ meant the very definition of intelligence quotient, IQ, given by the German psychologist William Stern at the dawn of the twentieth century (Gould 1996, 180). Though modern approaches diverge from this definition considerably, the characteristics of IQ scales still remained controversial. I diverged a bit from Suppes' account, however, not the least because I find his using the term "IQ" for intellectual ability as distinct from $S(a) / A(a)$, in other words, the very definition of IQ, confusing.

16. It is worth observing that order scale for the $iqa = S(a) / A(a)$ quotients could have been provided by a stronger premise. Namely, if we stipulate that S for the scores instantiates an interval scale. Indeed, it *is* generally stipulated by the students of intelligence (e.g., the common "calibration" of an IQ scale requires standard deviation—a statistical maneuver without any meaning on an order scale—despite the lack of any empirical evidence that any kinds of data on abilities would exhibit an interval scale).

17. In fact, Mohs' scale renders samples not scratching each other into the same hardness class. In this setup we may well find that some samples will belong to different classes at the same time—again a tricky traverse in front of linearity.

18. Other major works on the topic were written, e.g., by Pfanzagl, Roberts, and Narens (Pfanzagl 1971, Roberts 1985, Narens 2007).

19. After the testimony of Michell (Michell 2007).

20. Note that this axiom, hardly by chance, highly resembles the so-called Dedekind cut, a method for constructing real numbers (Dedekind 1901).

21. As mentioned before, the endeavor of building temperature scales on mixing did not prove to be fruitful in practice (Chang 2004, 60–64).

22. It comes as no surprise that the range of representation for these structures contains ordered pairs.

23. Humidity can be understood in different ways: as a relative quantity related to the dew point of the air on a given temperature expressed in percentage or as an absolute quantity expressed in, e.g., gram/liter. I am not concerned with this difference here.

24. Ignore the issue that the scale of temperature (and maybe even that of humidity) must be bounded in some natural way in this measurement situation.

25. I will not go into technical details here. A nice description of the practical applicability of the principle of conjoint measurement with many examples from behavior sciences can be found in Michell (Michell 1990).

26. The underlying logic for some theories may go well beyond the classical one. For instance, though Andréka, Madarász, and Németi strongly argue for first order logic, they use a many sorted version, and even Henkin-style second order logic (Andréka, Madarász, and Németi 2002). Hannan, Pólos, and Carroll developed a sophisticated multi-modal logic for handling theories in sociology (Hannan, Pólos, and Carroll 2007). (Indeed, my arguments in this book on the constructive nature of truth also enforce a non-classic, i.e., intuitionist logic.)

27. Thus Hempel's paradox, later indeed popularized by Hempel, is a nice example of Stigler's law: nothing is named after its inventor. It worth noting that according to Stigler's testimony, Stigler's law is also an example of Stigler's law, being invented by the sociologist Robert K. Merton (Stigler 1980).

28. See Hilary Putnam's foreword for Goodman's book (Goodman 1955, vii–xvi).

29. This is what Laudan calls *deductive underdetermination* (Laudan 1990a).

30. Just as Popper famously insisted (Popper 1935).

31. To be sure, there may be reasonable considerations what to choose from the consequences of a theory to test. See, e.g., Laudan's book on this (Laudan 1990b).

Chapter 4

On the Hard Road of Practice

If we are to sum up the accomplishments of the axiomatic project, we can say that the approach likewise encompasses good insights from the classical and the operational accounts. There is a broad family of scales rightly to be called measurement, but the given structure is to be empirically exhibited. It is a task of the practicing scientist to find out what is the case for a specific attribute. The issue is not logically necessary in any case, we work only with hypotheses always subject to empirical test. The axiomatic results also provide the scientist with clear methods to accomplish her task: for example, she can examine if a quality satisfies the conditions to be regarded as an additive quantity or not by the application of conjoint measurement.

This sounds like a success story. All the more surprising, there seems to be a consensus on the fact that axiomatic measurement theory never fulfilled its promises, never took the place in the practice of science which it would deserve. In particular, it never induced the hoped comprehensive changes in the methodology of social sciences, the "revolution" did not occur (Cliff 1992). It is not to say that the approach left science totally intact. Roberts devotes a whole book for the applications of additive structures only (Roberts 1985). Kahneman and Tversky fruitfully applied the axiomatic approach in decision theory (Kahneman and Tversky 1979). It had major impact on several projects in psychometrics, such as color theory or magnitude estimations, and also on some issues in sociology such as merge of ratings (Narens and Luce 1993). True, these latter studies do not necessarily live on detailed axiomatic analysis. But, for sure, they are keen on apt methodology and meaningfulness. Still, the least we can say that the success is partial, the expected breakthrough failed to come. Today social sciences still tend to follow somewhat opportunistic practices when it comes to measurement, and their living on the textbook-invigorated Stevensian sentiments is rather the rule than the exception.

In this chapter we will go through some of the explanations of how such a vast intellectual contribution could have passed almost unrecognized. We will take a glance at the practice of sociology, and we will tarry a bit with conceptualization and operationalization as key issues. Again, we will look into some aspects of preference measurement as a kind of minor case study. Theses observations will suggest that the exhibition of congruent phenomena may be the motor for well-grounded measurement procedures in social sciences—just like in physics.

4.1 ANATOMY OF A NON-REVOLUTION

It may be no exaggeration to say that Joel Michell (Michell 1999, Michell 2007) drafts a bill against the methodological practice of psychology (and in particular psychometrics) when it comes to measurement, describing it as "pathological" (Trendler 2009, 579).[1] He finds that not only the quantitativeness of a psychological concept is almost never tested, but also this negligence is disguised by a misguided measurement concept—the Stevensian one. The way out would be to use the results of the axiomatic approach (e.g., conjoint measurement) to show the quantitativeness of the observed quality at hand. However, psychologists (and, in general, social scientists) have failed to realize the revolutionary significance of these theoretical developments for their practice—at least so far.

As we have seen, the axiomatic approach—built on solid mathematical grounds—suggests that Stevens' initiations on the problem of measurement are valid. Nevertheless, his definition dissolves the concept of measurement, and is quite apt for veiling methodological negligence. According to Michell, it is more than problematic that a Stevensian measurement concept is still ruling in the realm of psychology and social sciences despite the fact that a solid conceptual foundation was offered since that, and we already have tools for deciding on the quantitativeness of certain qualities.[2] In itself, it is not an error to stipulate that certain qualities are quantitative, but forgetting the need for the verification of the hypothesis definitely is. A sorrowful fact: this error has already been integrated into the methodology of social sciences. Sure, errors are frequent in the history of science, but such a great carrier for an erroneous methodological approach may not be common.

The explanation for the spectacular failure of the error correction function of science, says Michell, is that the representatives of social sciences—let it, unconsciously—added a new error to the original one: did everything to veil it. Instead of facing the unjustified hypotheses, they advertise their quantitative discipline under the shields of the Stevensian definition.

It is not the case, however, that Stevens' theory on the levels of measurement and his definition of measurement are bound together for good and all by some logical constraints. The first does not infer the latter. But those who accept these as a two-in-one solution will not bother about justifying the quantitativeness of a certain attribute anymore. Often they simply stipulate that a given body of data is on, say, an interval or a ratio scale, and finding that this stipulation leads to more or less reliable predictions, they immediately assume that their ways are justified. True, Stevens' definition is rarely interpreted so that really any assignment of numbers can be regarded as measurement (e.g., random assignment is often excluded), but many insist that once somehow one got in possession of a rule, she does measure.

Though we have seen that Stevens definitely draws the lines for the permitted statistical tools for a given level of measurement, he is seemingly rather permissive with "illegal statistizing." He maintains that it is indeed not so nice to use disputable tools, still, these methodological frivolities may lead to fruitful findings, thus they can surely be forgiven (Stevens 1946). He was self-consistent in his inconsistencies: in line with his methodological credo, he would show some "liberality" in his research himself. It seems today many social scientists regard their trespasses so venial that they do not even see them as trespasses at all.

According to Michell, the Stevensian definition not only cuts the connection between measurement concept of quantitative sciences and mainstream psychology, but blindfolds those who accept it. These scientists ignore the need for the justification of quantitativeness. The Stevensian concept with its inconsistencies blurs the methodological problem of measurement and so to say "rationalizes scientific negligence" (Michell 1999, 20).

Michell has a point when bringing the charge of negligence against the practicing psychologists and social scientists: it is really widespread to regard measurable whatever object, attribute, phenomenon at hand, simply by maintaining that measurement is nothing more than the assignment of numbers. And once we have the numbers, we can add, divide, average them however we like. True, there are efforts being made on the verification of the quantitativeness of the attributes in many projects, but it is by no means widespread. All in all, few are keen on the tiring methodological issues of measurement in the ever-growing society of behavior and social scientists.[3]

While mainstream psychology (and, as it matters, *other* social sciences—if are prone to accept a liberal classification) have been living with the definition of Stevens for more than a half of a century now, a "revolution" occurred: the development of axiomatic theory for measurement. What is the reason, we could ask along with Michell, that the social sciences are still reluctant to acknowledge this revolution?

Several answers are possible. Some say that the applied mathematical language (the language of set theory) is unfamiliar to the psychologist or that representatives of mainstream psychology apply entirely different methods in their fields as the originators and supporters of axiomatic measure theory on their owns (Cliff 1992, 188). It is sometimes also mentioned that the useful applications of the theory have little relevance for other fields or that even this approach does not offer remedy for a major theoretical problem: measurement errors (ibid., 188–189).[4]

As for me, I think that the reasons for the durability and hegemony of the Stevensian views lie, at least partly, in their productivity. Clear, it is easier to arrive at results (or maybe "results") with a less strict methodology, provided that this less strict methodology is standard on the given field of study, and thus the results are accepted by the scientific community. Methodological scrutiny is time consuming, and eventually not rewarding, for the knowledge gained by hard work is though more well-founded, it is surely less "extent" than the knowledge gained by a sloppier methodology. Though it is a valuable asset to have research projects on solid grounds, this tiring activity may often result in weakening theories, may bring non-sequiturs or meaninglessness to the surface. In short, measurement theories may often turn out to have destructive rather than fertilizing effects on research projects, in contrast to, for example, statistics. As there is a competition among research projects for resources as in productivity, the situation does not motivate embracing stricter methodological principles. It is not an easy mission to persuade someone to change her comfortable ways to a tiresome track with less promise for fruits.[5]

Still, Michell insists, social sciences should change their ways. Social scientists must abandon scientism, that is, they should not try to prove all the time that they are able to provide quantitative result just the same as physics. They should admit that the applicability of a measurement concept like the one of physics is limited in social disciplines. In return, they should be proud of their many non-quantitative methods, where the applicability of the quantitative ones cannot be shown. The world is complicated, there are many complex attributes and it is not hard-coded anywhere that quantitative methods should always be relied on.

Michell goes so far as to ask for intellectual righteousness. He finds that the institutionalized practice of social sciences stands in the way of critical methodological investigations—a fact undermining their mere claim for a title as intellectual enterprise. Social factors drove social sciences to a point where critical methodological questions cannot be addressed anymore. Scientific practice should rely on empirical testing of hypotheses, we should not do with simply regarding those hypotheses true without question in which our interest lies.

Though I would say Michell is by and large right, we should take his verdict with a bit of skepticism. Social scientists devote considerable time for methodological speculations. What if it is really not clear how to integrate measurement theories in their methodological ways? It is hardly a surprising statement that when it comes to quantitativeness, social sciences readily turn to the vast and rich armor of statistics. It is also well-know, however, at least since Huff (Huff 1954), that these weapons are often blunt or even much too sharp but awkwardly manageable, easily misusable and sometimes self-dangerous. We can see an urge even in Stevens' work, however halfhearted, for taming statistics with proper considerations on scale types.

Nevertheless, the relationship between measurement theories and statistics is a bit complicated. Narens and Luce describe their affair by the following contrast (for the sake of simplicity): while "statistics focuses mostly on randomness, largely taking structure among variables for granted; [axiomatic measurement theory] focuses almost exclusively on structure, largely ignoring randomness" (Narens and Luce 1993, 129). Thus statistics and measurement theory represent two different approaches to the same problem: the challenge of uncovering "structure among variables in the presence of inherent randomness" (ibid., 129).

Social scientists willingly use highly sophisticated statistical methods to find some order in vast and inherently random data. Factor analysis,[6] as an example, is a popular tool for exhibiting patterns in huge, messy datasets. But once the proper factors are found, it is often *stipulated* that they instantiate an interval scale, without any real investigation on the possible structure. It is "taken for granted." Generally, this assumption is simply untenable.[7]

On the other hand, the measurement theorist assumes that the data "satisfy in one or more empirical interpretations" (ibid., 129), and tries to find some kind of numerical structure by which data can be represented. The axiomatic description does not take notice of randomness and errors, rather it seems to reflect some ideal state of affairs. But real data are seldom ideal. One could ask for the help of a possible error theory, but as we have seen, the axiomatic approach is not so ardent in this regard. In addition it is to be noted that a given bunch of raw data has an epistemologically superior status than processed data; it is "what is there," it is our first foothold providing a more or less direct link to nature (however we understand the notion). That is why one should be very careful in "taming" data.

Bridges, as Cliff puts, are hard to build for many reasons (Cliff 1992). Such is the "language" of the applied mathematics, the diverse methodology of social research projects, and the restrictive effect of methodological strictness on productivity. But even if major linking works existed today, it would take a long time to see their effects in the methods of a new generation of social scientists.

But there may be even more cloudy outlooks. Some say that measurement, properly understood, is simply impossible in social sciences. According to Trendler, an attribute is measurable if it satisfies the conditions for being a quantity, exactly as the paradigmatic physical attributes, length, weight and so on do (Trendler 2009). The only question we can meaningfully ask about psychological (or other) attributes whether they can be measured in this sense. For Trendler, this sense is certainly stronger than that of axiomatic measurement theory, since it brings back the concept of quantity of the empiricist approach. However, as I hope to have shown, finding a solid foundation for these "classical" quantities is neither self-evident nor unproblematic at all.

Trendler recites Michell and his warning that to decide whether an attribute has a quantitative structure is an empirical question—it is never logically necessary (Michell 1999). The *scientific task* of verifying a quantitative structure is rarely trivial, because the observable relations between objects (e.g., one piece of marble balances the other) and the relation between magnitudes (their equal weight) are logically distinct (ibid., 70). That is why, "we cannot take for granted that equal levels of some manifest variable necessarily correspond to equal levels of some latent variable, but we must ascertain *by experiment* that this really is the case" (Trendler 2009, 584, emphasis mine).

As Trendler rightly emphasizes, there is also an *instrumental task*: to develop measuring instruments—and according to him this is always secondary to the scientific task (ibid). I would inclined to say that the two tasks should go hand in hand: how would one establish data on relations without the appropriate instruments? Trendler writes, quoting Michell, "The first and therefore most basic condition of quantity structure demands that 'any two magnitudes of the same quantity are either identical or different'" (Michell 1999, 52, quoted in Trendler 2009, 582). Now, simply identifying equal levels of an attribute in different objects may require sophisticated instruments (often, but not necessarily based on sophisticated scientific theories).

Of course, establishing equality is just a first step. We also have to verify equal differences and possibly a meaningful concatenation concept (being there any), exactly as the empiricist account of measurement requires via its rules of measurement. And all there is at hand for all this work is experiment, nothing else, Trendler would say.

There are two major tasks an experiment must deal with. First, in an experiment we must independently manipulate agents to see the manipulation's effect on the observed phenomenon. Second, the experiment must hold disturbances, that is, agents and phenomena that are not objects of the given observation, under its control.[8] For all this we need a (most probably artificial) *apparatus*, which allows for independent manipulation and isolation.

The view that science should progress by actively intervening into natural processes (through experiments) rather than passive observation is called the Galilean revolution by Trendler (ibid., 587).

It comes as no surprise: "if psychological phenomena are not dependent or cannot be made to depend on a manageable set of conditions, then they are not measurable" (ibid., 590). No doubt, it is true that psychological (and social) phenomena are responsive to certain kinds of manipulation. It is also a fact that psychology does conduct experiments. But this all is hardly enough to show quantitativeness. Generally, it is already a problem to establish equal levels of a psychological quality.

Trendler concludes:

> *[P]sychological phenomena are not sufficiently manageable. That is, they are neither manipulable nor are they controllable to the extent necessary for an empirically meaningful application of measurement theory. Hence they are not measurable.* In my view no substantial progress will be reached in psychology until we accept psychological phenomena as they really are, namely in their natural "muddled" state. It might be cold comfort, but physicists would find themselves in the same hopeless situation if they were not to be allowed to construct apparatus. (ibid., 592) (Emphasis original.)

The (Galilean) revolution never happened, and cannot happen in psychology. Psychologists (and in general social scientists) must find other, non-quantitative methods to achieve progress in their fields.

It is no wonder if this conclusion sounds somewhat familiar. Indeed, it is very similar to that of the Ferguson committee in the thirties (see section 3.1). The committee's manifestation triggered Stevens' operationalist answer thereafter. Then, it remained to the axiomatic theorists to make good use of Stevens' ideas in building a strong theory with mathematical rigor. Ferguson psychologists and also Stevens were accused of dissolving the meaning of measurement by using new and sloppy concepts of it. This criticism was, to a great extent, legitimate. But axiomatic theorists did a lot for clarity and conceptual soundness. Reverting to an eighty-year conclusion after their work does not seem to be right.

Moreover, we well may have reason to not be content with Trendler's concept of experiment. In particular, what is *sufficiently manageable* is not clear-cut. It is to be observed that natural science experiments are also full of unmanageable elements and tacit presuppositions. What guarantees, anyway, that the standard prototype unit or the behavior of my measuring instruments stays sufficiently similar between t_1 and t_2? What is sufficiently similar? On the other hand, as we have seen, Regnault sufficiently managed his thermometers without any elaborated theoretical background. Such reliance on congruent phenomena is open for social sciences, too.

What we see here is the same old story of different concepts. Trendler regards the empiricist measurement concept as the only legitimate one. According to this, the rules of measurement are genuine scientific laws as conceived by natural sciences. I intended to show that this view is not easily tenable. The measurement concept of the axiomatic approach departs from this, no doubt. But why not? Concepts are not fixed for good and all. Transcending them is sometimes called progress.

Talking about the limited success of axiomatic theories, I would add some further points to the list above. First, we could ask what does exactly the axiomatic measurement concept encompass? Where are its boundaries; what "counts as" measurement and what not? Second, how "telling" a representation is, what is its real added value to know the exact structure of data?

The notion of measurement is, of course, vague—just as almost every concept.[9] It is vague even under an axiomatic account. Here, the concept of measurement is characterized by a theory (an axiom system using a formal language) and its model's homomorphisms. Having feasible empirical interpretations of the primitives in our language, the models instantiating our system are homomorphic to some numerical structure. But what do we mean by *numerical* (an epitheton ornans in the axiomatic literature), and why do we need this restriction at all (if it is a restriction)? Do we regard partial ordered sets as "numerical"? Hence, is establishing some partial empirical relation or possibly some quite weak operation a measurement? Again, derived and conjoint structures involve models with ordered pairs (or n-tuples). Are they "numerical"?

Assume that we can drop the attribute "numerical," and we can freely talk about simply "mathematical" structures instead, in other words, abstract set theoretical structures—which may well be in line with the intention of the students of axiomatic theory.[10] Suppose further that we have some dataset with complex, but axiomatically describable features. Now, the models of our theory will not be homomorphic to any of the structures usually associated with numbers. Will the involved empirical procedures count as measurement?

Consider the theory of complex networks as worked out by Albert and Barabási (Albert and Barabási 2002). The authors analyzed many different networks from the World Wide Web through cellular and phone call networks to citations and Hollywood actor collaborations. Through examining large datasets they had to "measure" these networks, that is, account for "nodes" and connection between nodes to exhibit their topological features. That is how they found some interesting common features of them: for instance, they are all small worlds (in the sense that there is always a relatively short path between node pairs), scale-free (i.e., follow power-law distribution, which in practice means that they exhibit relatively few but rather heavily connected "hubs"), and so forth.

Now it is tempting to set up an axiom system with proper empirical interpretations on these datasets. It might not come as a surprise to us that the models of our system will show homomorphisms to certain graph-theoretical models. But we may acknowledge this fact with a tiny bit of discomfort. The thing is that graph-theoretical considerations are deeply involved in the very theory; all of its major findings are graph-theoretical assertions.

To make it clear, I would willingly admit the above situation as measurement. But now the question is, what would we gain with an axiom system for this kind of measurement? One could readily answer that we will have a warrant that we made no mistake and our empirical procedures are really reflecting the intended structure. But honestly, we would be a bit surprised if it did not, and we might even be prone to blame the axiom system.

While axiomatic measurement theory may be immensely useful and fertile in certain cases, it may be dumb and annoying in others, just like some customs officer claiming for a third copy of each certificate. No wonder that some scholars find the whole project inefficient. As Zoltan Domotor put sarcastically: axiomatic measurement theory is "the enterprise of poliferating boring corollaries to Hölder's theorem" (Michell 1999, 198).

This verdict is without doubt excessive and unjust. Nevertheless, even occasional impressions of triviality may make the axiomatic theory unappealing and may well add to reasons of its lack of real success.

4.2 CONCEPTS AND DEFINITIONS

Standard textbooks in sociology often claim that the measurement of social phenomena is essentially the same as those of physics, only the results are less exact because of the very nature of the attributes measured (Steele and Price 2007). Sometimes they go as far as to say, "researchers can measure anything that exists" (Babbie 2007, 121). There is often a warning that the reliability of our measuring instruments should be tested and measurement procedures should be verified as valid, but after all it is simply presupposed that certain attributes can be quantified. Everyone, however, taking a closer look on some quantitative results of sociology must come to the conclusion that the discipline is immensely affected by cruxes set forth against psychology by Michell. Stevensian sentiments are alive and well in sociology, undisturbed by the enormous developments in measurement theory. The textbooks mentioned often quote the theory of measurement levels uncritically and without further comments on the subsequent developments (ibid., 136–140). Every now and then we bump into definitions which make clear that measuring is simply an assignment without the obligation of empirical verification:

"[m]easurement is any process by which a value is assigned to the level or state of some quality of an object of study" (Bulmer 2001, 455).

Clearly, sociologists often deal with concepts, such as contentment, the quantitativeness of which is at least rather doubtful (but could be tested). But many social research projects rely entirely on data where only counting is involved: a measurement determines how many people belong to a set such and such. Data so gained are on an absolute scale in Stevensian terms, hence there is free way for meaningful statistics. What is more, the notion of unit is even clearer than what a physicist may ever dream of! But the thing is that sociologists are rarely after pure numbers, they tend to measure complex (allegedly) quantitative concepts through the data. Concepts which are poorly defined or even yet to be understood.

It would be unjust, however, to maintain that sociologists deliberately and assertively neglect the rules of scientific integrity and ignores the need for critical inquiries. It is often the case that they spend a lot of time on construing proper methods, and even on bewailing the contingency, imperfection, deceptiveness of them. They often tend to be rather critical and doubtful regarding their ways, this is well indicated by the many research projects devoted to methodological pluralism, that is, the parallel application of different methods for approaching the goal of a scientific project.

On the other hand, social research often looks for complex, multidimensional patterns of correlations and identifies underlying factors. While using factor analysis and its broad kinship of methods, sociologists often arrive at scales with stipulated properties (such as equal intervals) and draw bold consequences. Adding to this, they too often *reify* (Gould 1996): they regard the derived factors as they were real, existing entities in the world. Of course, picked out factors say little without interpretation. This is by no means to say that these methods of analysis do not convey any *information* whatsoever, only that one should be much more careful with inferences. Of course, this precaution would lead to more modest results. As I pointed out above, this feature in itself is a significant drive for loose methodology.

Now, to dive into the main concerns of measuring the social, let us take an everyday measurement situation as sketched by Taylor (Taylor 1997, 3–4): a carpenter plans to install a door, and to do this, first he needs to know the height of the doorway. While improving his methods from pure-eye estimation to using a measuring tape and finally a laser interferometer he gradually boosts his precision, or reduces the margin of error—eventually to the order of the length of the light wave. But however he would improve his precision, measurement errors would remain. Their order might be practically neglectful related to the observed magnitudes, but their nature would remain the same.

But this is not our main concern now. There is also a serious conceptual problem: what do we mean by the *height of the doorway*? One may say that the perpendicular distance between the floor and the ceiling. But perpendicular to what? Measuring perpendicular to the floor may well result in different values than measuring perpendicular to the ceiling. And what do we mean by perpendicular on a by all means imperfect surface? One might answer, ok, let us measure *vertically*! But now our problem is to precisely measure the direction of the center of the Earth. Moreover, removing a thin layer of dust our measurements may well yield different results than before (with a tool of appropriate sensitivity). Even more importantly, we may yield different results by measuring the height in different points. One might answer this time that we can specify exact points for our measurement by their distances from the walls. But what do we mean by the *distance from a wall*? That is, all that we do now is reducing the conceptual issues of a given measuring task to a different but likewise vague measuring tasks. We can call this the problem of definitions, or the problem of fuzzy concepts.

In formal studies, "well-behaving" concepts introduced by sharp definitions are abundant. The set-theoretical definition of ordered pair, for example, is clear and unambiguous: we can always decide whether a given set is an ordered pair or not by its mere form:

$$(a,b) =_{df} \{\{a\},\{a,b\}\}.$$

All right, one might say, but it is just one definition of the pair, known as Kuratowski definition. There are several others. Here is Robert Wiener's, for instance:

$$(a,b) =_{df} \{\{\{a\},\emptyset\},\{\{b\}\}\}.$$

Is it not just the case, then, that there is no strict and generally accepted definition for an ordered pair, so the notion is as vague as any other? By no means. It can be admitted that the different definitions characterize different ordered pair concepts, and the so defined pairs show diverge properties, which may have serious consequences on the whole formal system we work in. The concept of ordered pair can be said to be *ambiguous* until the point we swore on one construction. Once a definition is chosen in a given context, the notion is fixed for good, there is no way for switching to an other notion. Also, there are no borderline cases: what an ordered pair is and what is not is clear-cut.

Outside formal systems, definitions tend to go less tame. What is length? We may have several suggestions: length is the most extended dimension of an object; length is any quantity with a dimension distance; length is the *measured* dimension of an object; length is the linear extent of an object in space from end to end, and so on. Now we might insist that the concept of length is just as ambiguous as that of the ordered pair, but fixing an appropriate definition for our context will provide a strict notion. This is not so, these definitions are vague, too. Some difficulties stemming from using likewise vague concepts in the definitions ("dimension," "distance") may be overcome by giving sharp definitions for those ideas. But, of course, the danger of other vague notions occurring in those very definitions still lingers on. Moreover, some concepts cannot be sharpened out, even in theory. In order to know what is the most extended dimension of an object, we should measure it. But the difference of the observed magnitudes may well lie beyond our margins of precision. Of course, this concept is particularly unfortunate, because it involves the very notion of length measurement. But similar consideration are in place with regard to the *end* of an object or even to its *linear* extent.

Again, suppose we settled on a notion of length, which seems to be apt and strict beyond all possibilities. Now we face further issues when dealing with concepts like a rod which is one meter long. If we are asked to sort out from a given set of rods the one-meter-long ones, we are able to do this task with some approximations. But be our margins of precision in whatever order, there will always be *borderline cases*—rods that are almost-one-meter-long or a-tiny-bit-more-than-one-meter-long, but they can as well be considered as one meter long with regard to our margins of precision or practical purposes. This uncertainty is immanent and inevitable: there is no way whatsoever to sort out the rods which are *exactly* one meter long.

A widespread strategy for sharpening theoretical concepts is the deployment of *operational definitions*. In general, this way we can define entities by describing how to exhibit, produce, or create them. In science, by an operational definition we usually and broadly mean a description of a measurement procedure—as we have seen, for operationalism measurement is a central notion (see section 1.4). Operational definitions can well serve science and practice by making blurred concepts sharp and meaningful. Thus, quantum mechanics defines field properties by the measurable properties of observable particles. Relativity theory defines distance and simultaneity by their respective procedures of establishing them. And IQ can also be, and often is, defined by the score achieved on this and that mental test.

In social sciences, operational definitions are abundant in their trying to make their concepts quantifiable. Properties of living standards, corruption, migration are hard to measure from the start. When finding a viable way for a quantitative account, the original meaning[11] of the concept narrows. It is clear that meaning cannot be fully determined even in natural sciences. The situation seems to be less chaotic there still, because there is much more room for law-based, systematic measurement (Kyburg 1984), or measurement in experimental setup (Trendler 2009). These conditions put constraints on the possible measurement operations.

Let us take the relatively simple example of transport safety. In the table on figure 4.1, I made up some possible operational definitions for transport

$TS_1 =_{df} \frac{\text{Injuries}}{\text{Mileage}}$	$TS_2 =_{df} \frac{\text{Injuries}}{\text{Traveled hours}}$	$TS_3 =_{df} \frac{\text{Injuries}}{\text{Number of people transported}}$
$TS_4 =_{df} \frac{\text{Fatalities}}{\text{Mileage}}$	$TS_5 =_{df} \frac{\text{Fatalities}}{\text{Traveled hours}}$	$TS_6 =_{df} \frac{\text{Fatalities}}{\text{Number of people transported}}$

Figure 4.1 Definitions for Transport Safety.

safety, *TS*, where, of course, the safer ways of transportation score lower. This list is by no means complete: it is enough to consider what is to be counted as injury. It can be any harm more sever than a bruise, or any harm more severe than a broken leg. But this simplified picture is enough for making some observations.

First it is worth making clear that these definitions do not exhaust the meaning of transport safety—where by meaning, I mean how this phrase is used by experts on a conference dedicated to the topic. And there is a good reason to maintain that we cannot even cover it in theory would we be able to create as many operational definition as we like. (And we can easily create infinitely many, as figure 4.2 shows.) Not the least because these are certain aspects of this use, which escape from a quantitative approach.

A social scientist with naturalist inclination, by which I mean she is keen on applying the methods of natural sciences as the only legitimate ones, may

$TS_7 =_{df} \frac{\text{Damages over \$1000}}{\text{Mileage}}$	$TS_8 =_{df} \frac{\text{Damages over \$1000}}{\text{Traveled hours}}$	$TS_9 =_{df} \frac{\text{Damages over \$1000}}{\text{Tonnage transported}}$
$TS_{10} =_{df} \frac{\text{Damages over \$1100}}{\text{Mileage}}$	$TS_{11} =_{df} \frac{\text{Damages over \$1100}}{\text{Traveled hours}}$	$TS_{12} =_{df} \frac{\text{Damages over \$1100}}{\text{Tonnage transported}}$
$TS_{13} =_{df} \frac{\text{Damages over \$1200}}{\text{Mileage}}$	$TS_{14} =_{df} \frac{\text{Damages over \$1200}}{\text{Traveled hours}}$	$TS_{15} =_{df} \frac{\text{Damages over \$1200}}{\text{Tonnage transported}}$
...	...	...

Figure 4.2 Some More Definitions for Transport Safety.

insist that the meaning of a concept can only be given by an appropriate quantitative definition. Anything beyond it does not constitute any meaning at all, it is just empty chatter. So far so good, but now she must take the inconvenient task of choosing one of the possible definitions. The problem is that all of the definitions are equally legitimate, there is no theoretical reason to choose one above the others (though there may be practical reasons: e.g., a choice could be motivated by the available datasets). To be sure, measuring along different operational definitions may yield quite different results for such simple (true, not really exact or well-formed) questions as: "what is the safest way of transport?"

A possible way for giving the issue a better outlook is to say that, though operational definitions cannot exhaust the meaning of a concept, they can fix it *in a given context*. Another context may involve different but equally legitimate notions. What ties these concepts together more strongly that any two random concepts is a common *umbrella concept*. But these umbrellas, constituted by all the legitimate operationalizations, are not strict scientific concepts, would the (soft-hearted) naturalist say. Nevertheless they are part of the scientific practice: as catchphrases in the sloppy metalanguage of science.

4.3 CONCEPTS AND VARIABLES

It might be to the sociologist's delight that there is a vast number of numerical or trivially numerizable databases (or datasets) available. On these "hard data" we can measure simply by counting the number of occurrences, adding up (extracting, averaging, etc.) entries representing time (say, working hours), amounts or individuals.[12] Since these measurements are, so to say, on the absolute scale, their statistical treatment is not really problematic. As a trade-off, they raise other serious methodological problems.

Sociologists call the entries of these datasets *variables*. Some of these variables provide a pretty good base for carrying out measurements. However, in sociology even the simplest ones of these measurements are indirect. Say one can measure (estimate) the number of foreign citizens working in the country based on the data on citizenship in the database of the national health insurance system. Here is an immediate methodological issue: the exploitability of administrative data sources.[13] Data stored in, say, registers[14] are not for scientific purposes at the first place, they are not collected the way to provide with immediate information for the sociologist on her questions of interest. Measurement on them is not only indirect, but also based on more or less stern presuppositions on the idiosyncrasies of data collections. In the example above, we would like to know the number of the foreign citizen, but what we measure is the number of those foreigners who were registered and were not unregistered by their domestic employer up to a given date. Thus not only unregistered employment is hidden entirely, but we cannot even be sure how the our results are related to the "real" situation. Often, the data lack comprehensiveness and are not up to date—datasets exhibit wide variance in this regard. And, of course, talking about administrative data sources, the access to them for scientific purposes could also be highly problematic. Governmental data is often withheld in the name of personal data protection, but the real obstacle are rather maleficent routines and ignorance. It is not impossible to maintain data systems where the access for anonymous governmental data is unobstructed—Scandinavia is pioneering on this field ("Out of the Box" 2015). Nevertheless, secondary use of existing data is of crucial importance for social sciences. It is not by chance that recently several initiatives aim at the development of this important methodological tool ("PROMINSTAT Project" 2010, Csatári and Juhász 2009).

Beyond hard data, variables tend to become more awkward from a measurement point of view. Set now apart the entries of anecdotal nature one often comes across in surveys, they are part of the "soft," qualitative side of social sciences—by no means less valuable than the quantitative side, only we are concerned with the latter in this study. No measurement, no problem—at least from our perspective. Binary variables, like answer selection (yes/no), or (traditionally) gender can easily and legitimately be "statisticized" with, even though one may wonder how much dealing with these data fall in the usual sense of measurement. But other entries, such as ones coding preferences, should not (though usually are) regarded as quantitative, at least, not by default.

Say, we are to examine the contentment with the work of government on a ten-grade scale by asking respondents via phone. According to the practitioner's advice (Steele and Price 2007), if we received a "7" as an answer from John Johnson and we call him again the next day and he gives a "3," we should suspect that our "measuring instrument" is unreliable. Likewise, if we found on a large enough sample that the mean of the result was 9.8, the validity of our measurement should be questioned. Nevertheless, the main problem with preferences is not even touched in these warnings. Namely, we do not have any reason to suppose that John Johnson is more content with the work of the government than Kate Kenneth, who happened to give it a "5." And from the start: why to think that we will get the same answer from Louis Lark before and after lunch? This is not to say that a preference scale cannot be meaningful. Only it should be verified, possibly by some of the methods set forward by measurement theories or by successful iteration.

In any event, the least to say is that their variables provide the effective means for sociologists to approach the concepts in scope. Variables are generally conceived as operationalizations of concepts, which are in turn regarded as crucial in confirming causal theories by a traditional physicalist point of view.[15] It is important to see that variables as constituents for a procedure of measurement, in the view of practitioners, do not usually *define* a concept, as the old school of operationalism would suggest. Rather they may make up for some operational definitions at best, ones that more or less grab a blurry moving target, and which are not exclusive, only by chance chosen ones from the infinitely many possible. Of course, the exact way of operationalization often depends on the contingency of available data. As Bulmer puts, characterizing, for example, the health of a population or crime in a locality can be realized through a vast set of possible measures (Bulmer 2001).

But the proliferation of variables by no means helps conceptual clarity. Sartori writes:

> [M]uch of what is currently labelled social science "methodology" actually deals with research techniques and statistical processing. In moving from the qualitative to the quantitative science, concepts have been hastily resolved and dissolved into variables . . . [C]oncept formation is one thing and the construction of variables is another; and the better the concepts, the better the variables that can be derived from them. Conversely, the more the variable swallows the concept, the poorer our conceiving. (Sartori 1984a, 9–10)

Indeed, contemporary "research techniques and statistical processing" include multidimensional data-processing (factor analysis and its kinship), where even the variables (factors) are somewhat wanton. The analyst is "free" to choose what resultant factors she defines from the original ones (at the expense of losing information). Also, several equally justified mathematical solution may exist for the same problem.[16] Interpretation, conceptualization comes last, as if we placed the target where the arrow landed. By and large, practicioners even with a healthy methodological incline are not keen on conceptual tidiness, though there are laudable exceptions (Sartori 1970, Sartori 1984b, Gerring 1999, Collier and Mahon 1993, Brons 2005).

"Stable concepts and a shared understanding of categories" (Collier and Mahon 1993, 845) are of utmost importance for each scientific enterprise, and it is clear that sociology, or generally, social sciences lack this kind of accordance. Interestingly, as Sartori observes (Sartori 1970, Sartori 1984b), not only the restrictive nature of operalization is to be blamed. Applying concepts of old theoretical frames in new fields often results in the unconscious departure from the notions: conceptual *traveling*. This is often followed by conceptual *stretching*, the concept is "distorted" so to fit in.

At this point the reader ought to recall Bridgman's motives for his operationalist approach (Bridgman 1927). His main point was that we have no reason to apply the same concept for properties measured different ways. Length measured by a measuring tape is not the same concept as length measured by a laser distance meter. We can see his attitude as the most cautious one to prevent any unconscious and unjustified conceptual traveling. Chang took on these thoughts to a less restrictive, positive reading. In his view operationalist considerations may lead to legitimate and coherent extensions of concepts (Chang 2004). To this end, two conditions must be satisfied, Chang writes:

> *Conformity*. If the concept possesses any pre-existing meaning in the new domain, the new standard should conform to that meaning.
> *Overlap*. If the original standard and the new standard have an overlapping domain of application, they should yield measurement results that are consistent with each other. (ibid., 152)

All this is to mean that once we have well-founded standards on different domains (which, as we have seen, can be reached by picking out phenomena of congruent behavior), a unified quantitative concept could be established. Unquestionably, this is a practice in natural sciences. For instance, the theoretical concept of temperature still has little role in its measurement. Instead, the practical temperature scale exhibited as a patchwork of overlapping operational procedures ("International Temperature Scale of 1990" 1990). In principle, one might feel, this way is also open for social sciences to follow. And there may indeed be such areas. It is really hard to see, though, how these principles could be applied for concepts like transport safety. Divergent concepts with non-overlapping, what is more incomparable operalizations are the rule rather than the exception in these disciplines.

4.4 THE CASE OF PREFERENCES

Below I will analyze a popular genre of data production in social research practice. It will come as no surprise to see that par excellence social sciences have a common fate with psychology in that the axiomatic principles have not gained real ground. By and large measurement theoretical considerations play little role in their practice, and it is stunning that while some practitioners blame social sciences themselves for not having formally embedded strict concepts, they completely overlook the vast developments in the field of theorizing on measurement.[17] Instead, they insist that we arrive to measurement whenever we classify a set of units by quantitative variables. Indeed, as a peculiarity for social research, sometimes we can do this in all conscience. As stated above, in many cases social research possess data, which do not face serious problems from a scaling point of view. For instance, doing with counting after all, measuring on administrative datasets allows a good start.

But often at this point, the relation of the concept and its operationalization is problematic. Concepts tend to be "under-defined," or in other cases suffer proliferation, stretching, traveling by the very way of operalization. In many cases the relation between the concept as an object of inquiry and its operalization is entirely unclear and unexplained—or explained only by its presumed effectiveness or success.

Elsewhere I analyzed a prestigious reoccurring public opinion survey and found it painfully weak in conceptual clarity and in conclusiveness (Csatári 2016, 133–136). I was not after some destructive writing off then, my intention was to show that laying stress on methodological issues does make a difference. Below I take a somewhat more general approach while analyzing the practice of preference measurement in different attires.[18] Once again, if I happen to have strongly critical insights, my intention is just to clear the brushwood to find a viable path ahead, and, of course, to yield some more fuel for the final conclusions of this book.

We may suspect preferences everywhere in our daily acts. Preference, of course, is not a unitary concept rather a type, which may reveal itself in various guises and contexts. Some of them might possibly be approached through quantitative methods others surely not. For instance, I would hardly esteem Debbie's morning preferences for her daily dress measurable. In other cases it may be easier to identify certain factors lying behind choices or intentions. My shoe preferences might well depend on (conceived) durability and price. In this latter case the method of conjoint measurement might in principle help decide on its measurability. But below I will take a more clear-cut case of one-dimensional measures, where no underlying numerical factors are presupposed and the data is gained simply by collecting the answers to a single question.

Such is the rather popular and widespread method of Net Promoter Score (NPS), which is, professedly and roughly, supposed to measure the contentment or loyalty of employees or customers—by every means an instance of preference. NPS was introduced by the business strategist Frederick Reichheld (Reichheld 2003), and, not entirely trivially among social science methods, also honored as a registered trademark associated with a couple of commercial products. The method collects answer to the question: "How likely is it that you would recommend *x* to a friend or colleague?" The interviewee is supposed to pick an integer between 0 and 10 as an answer. Those who choose 9 or 10 are labeled as *promoters*, those who take 6 or below are called *detractors* and those, who bet on the two numbers between are the *passives*. The score is calculated as the difference of the percentage of the promoters and the percentage of the detractors, thus represented on a scale running from −100 to 100.

At first glance it is not so easy to tell what these numbers mean—nor at the second, actually. As probability is represented by a number on the closed interval of 0 and 1, the set of integers given as possible answers with a trivial projection to this latter seems to be a wise choice. Only no probability is anywhere on the horizon. On the most plausible interpretation, the interviewees confess some presumed probability of a potential act, which is some vague counterfactual at best. Moreover, there is no cogent reason whatsoever to numerically relate Debbie's choice with mine, let alone when coming close to each other. I see no independent ways to confirm the original "measurement," not the least because the very concept considerably lacks clarity. Only due distance may provide some weak reason to order the results. To put it in the language of levels of measurement, after aggregation we gain something less than data on an ordinal scale.

After all this, transforming data to the scale from −100 to 100 further blurs the picture. The chosen categories are entirely arbitrary, and probably the only motive for the idiosyncratic calculating procedure is the need for a nice-looking data distribution. But okay, we can suspect marketing considerations at work here instead of scientific ones. Understandably, it may be of real fancy to project red and green gauges to the audience.

But there is more to this. As the common wisdom goes, while natural sciences have the opportunity to ask nature by experiments, the social scientist is left with asking people on what she is after. Here is an archetypal example where nothing akin happens. The researcher is meant to measure loyalty (whatever it might be), but asks, mildly put, a not trivially related question.[19] According to Reichheld, operalization with the given question *is* motivated. It was chosen from a bunch of similar questions as the best predictor of actual customer behavior, measured by other methods. Lack of data, we may only believe it, and also that it is properly clarified what we mean by prediction. At the same time, the same indicator is regarded as a good measure of company growth (in most of the cases). But at this point I stop, unwilling to go into further proliferation of hazy concepts around NPS.

One can object that dissecting a popular tool in business management is unfair and bears little moral for social sciences proper. But this is not so. I chose this example because it is a simple, widely accepted and used standard. The method is meant to measure. It represents the main features of the wide family of similar data collection, such indicators are abundant in research with social science methodology. Our example also nicely shows their common weaknesses: the lack of empirical meaning and conceptual mess.

But all right, I will turn to an example of perhaps a tiny bit more scientific prestige in a minute. Before that, some more words on what it means to be a standard are in place. Bulmer rightfully observes that the well-known conceptual problems with measurement in social sciences constitute the main obstacles to the constructions of unified indicators (Bulmer 2001). Nevertheless, it seems that standard indicators often brought about, so to speak, spontaneously. But what is the warrant that such a standard is a cogent, valid operationalization? There is of course the presumed success in prediction. But the belief in predictive force is prone to be highly unjustified, let alone when we are able to identify the nature of the preferred (or not preferred) outcome rather vaguely, if at all. But a clear lack of flames does not prevent one feeling the heat if she really wants to. Nevertheless, apart from any arbitrariness and conceptual mess, the fact that a wanton operationalization becomes a standard could be regarded as good news. By producing reoccurring comparable data on longer run an indicator may provide basis for weaker or stronger implications. On the other hand, it may also turn out to be a petrifaction giving no way for other approaches.

As for such a petrifaction, let us consider party preference measurement, where one applying non-canonical methods is immediately suspected of hiding something with making her data incomparable—a surmise not always unjustified. The standard procedure of party preference measurement is originated from a polling method of George Gallup. It is probably the most simple, most popular operalization of the concept. Gallup introduced it in 1935, and, like NPS above, also supposed to quantify a preference concept by asking a simple, and by now classical question: "If the election were held today, who would get your vote?" Slightly modified versions of the questions are also used, such as in Hungary where it goes: "If the election were held next Sunday, who would get your vote?"

Practically speaking, the survey has a double role. First, it serves as the assessment of the actual popularity of certain parties (generally it is not comprehensive with regard to the existing political organizations in a country—at least on the evaluation level minor ones are cast up). Second, especially when there is indeed an election around the corner, it serves as prediction. A historical fact is that this method had become popular and, so to speak, a standard among pollsters because of its success in predicting certain elections results. Its reliability is far from being uniform or unquestionable, though. Not only has it gone astray many times with such predictions, but also its credibility may differ widely between different societies. For instance, in the recent past it tended to give quite accurate predictions in Germany, but was really untrustworthy in Hungary. Again, it is prone to give biased results in the hand of certain pollsters with different political inclinations.[20]

As above, and contrary to the sociologist's credo, pollsters (and their policymaker clients) do not ask what they are curious about. Why would they need to know the population's feelings about a counterfactual? Let alone that we have every reason to raise doubts about the clear and immediate connection between someone's wishes, plans, intentions, and actual behavior. Indeed of course, pollsters are after public political mood, and this is operationalized through a survey method with a question chosen from a host of possible ones. According to Bulmer, variables for a given concept can only be justified in pragmatic terms and common sense (ibid.). So far so good. But he also adds: the concerning concept is operationally defined by the mere words used in setting up variables. What words? It seems that the wording of the question in question does not define any meaningful or interesting concepts, or if it possibly does, it is not what is meant by the researcher. Rather, as far as I can see, the only viable interpretation of Bulmer is that an exact wording of the *surveying method* is the one which could bring us to the concept of party preference. Without it, having no direct relation to the concept in scope, the actual question might operationalize *anything*.

At this point we should stop to notice what makes the standard of party preference much more sound than NPS. First, the former has much stronger privileges from the start to qualify as a quantity by being trivially and meaningfully numeralizable. Both of the surveys are based on counting the votes for different answers, of course. Still, differences are crucial. NPS applies abstract entries without definite meaning—still presupposing that the interviewees have a common understanding of them—which are in some stipulated relation associated with the usual meaning of numerals. On the other hand, party choices are definite, meaningful and clear-cut, each possibility is discrete and logically independent from the other. The so yielded data are absolute, and are eventually turned into percentage quite legitimately.

Second, party preference measurement is based on a sophisticated sampling method. The sample, the group of people who are actually asked, are chosen so to exhibit a distribution which shows similarity to the whole population of voters in important respects, such as level of education, age, urban conditions. That is, the measurement procedure involves a theory, according to which different social groups have different propensities to vote for a given party. This theory may provide an important adjustment to the method, and greatly contribute to the soundness of the procedure. As, of course, it is empirically testable.

But are we not again in a bad circle? Since in order to measure the party preferences in the different cohorts, we need an established method for it. But alas, we are about to establish the very method at this point. The worry is deeply justified, and the consolation I can offer is similar to the one I proposed above in regard to some physical quantities: we can turn to phenomenal congruence to be our guide.[21]

I see two—not unrelated—roles for congruence. First, it is the empirical anchor in calibration, when the same quantity is supposed to be approached by an armature of different instruments. Having a procedure at hand, we can relate it to other procedures aiming at the same concept if we are lucky enough to find such. This comparison may either be non-theoretical (or relying on a minimal theory) as in the case of Regnault's thermometers, or may contain sophisticated theoretical speculations as in the case of timekeeping.

Second, congruence can be a judge for the validity of a given procedure, either by iterative application or by giving identical "instruments" in different "hands." Thus, the standard method for party preference could validate itself if it showed similar enough results in the hand of different pollsters (or if it could be conducted frequently enough with cogent results). Of course, we can measure only once at a given point in space and time (so to speak), but it is legitimate to expect reasonably close results in case of reasonably close measurement instances.

All in all, though trying to make advances through sophisticated calibration, the immense discrepancies of results in practice undermine the classification of the Gallup method as a valid way of measuring party preference. The approach is promising, but its inertia may block further development. All the more sad, because the way forward could only lead through adamant quest for promising patterns. Lacking law-based measurement in general, social sciences should still systematically look into ways of justified standardization through congruence and meaningful treatment of concepts.

NOTES

1. Michell is a psychologist himself, but it is not a surprise to see a scholar to be highly critic, or even inimical to the methodological practice of his own discipline. A resounding example is Andreski (Andreski 1972).

2. Despite his appreciation for the axiomatic project, to Michell measurability and quantitativeness are characterized by the classical notion of Maxwell: measurement is the estimation of a magnitude by an arbitrary unit of the given quantity (Maxwell 1890).

3. As Andreski put it sarcastically decades ago: "To judge by quantity, the social sciences are going through a period of unprecedented progress: with congresses and conferences mushrooming, printed matter piling up, the number of professionals increasing at such a rate that, unless arrested, it would overtake the population of the globe within a few hundred years" (Andreski 1972, 11).

4. Indeed, the question of errors is not a central topic for the measurement theorist—and this is an admitted weakness of the project. However, error theories exist, e.g., one is outlined by Kyburg (Kyburg 1984), his theory will be discussed in the next chapter.

5. How much is the matter different in natural sciences or engineering and why—these are legitimate questions. Of course, a more thorough analysis would be in place to cover the board scale of cases in different disciplines from pure science to policy-driven (and founded) surveys. But this is not in the scope of the present text.

6. For an introduction to factor analysis, see the works of Gould or Kootstra (Gould 1996, Kootstra 2004).

7. Without going into technical details, it is to be required for a main factor to be on an interval scale that all of the great bunch of correlations involved in the analysis are linear. Ask yourself: what are the chances for one single manifestly linear correlation?

8. Trendler follows Maxwell (Maxwell 1890) in his account of experiments (Trendler 2009, 585).

9. Possible exceptions may be concepts fixed for the sake of a given "formal game."

10. A clear indication is that the second volume of *Foundations* (Suppes et al. 1989) is largely devoted to *geometrical* representations.

11. By meaning here, I mean informally the scope a given notion has in expert discussions.

12. It is not always trivial to maintain a "counting science." Consider biology. When assessing the extent of a population we may ask: how many individual corals are there on the reef? And it is not easy to find any meaningful answer (Godfrey-Smith 2014, 67). I see counting in sociology unproblematic, at least, in this regard.

13. The use of existing datasets for an unforeseen purpose is not without example in natural sciences. For instance, a recent study calculates ocean temperatures through data on atmospheric oxygen and carbon-dioxide levels of a more than two-decade collection (Resplandy et al. 2018).

14. One possible grouping of datasets is by their purpose, another by their method of collection. By the first dimension we can broadly distinguish between administrative and primarily scientific data collections. Not unrelated, by the second dimension, an apt choice is to talk about registers, counts, and surveys. (A taxonomy used in, e.g., PROMINSTAT, an endeavor for summoning migration-relevant databases in Europe ["PROMINSTAT Project" 2010].)

Administrative databases are maintained by the state or some local authorities for, no surprise, bureaucratic, or sometimes policy purposes. While other data are collected mainly of scientific interest, we cannot draw sharp margins, not in the least because scientific projects often serve policy purposes.

Registers are administrative databases encompassing information linked to individuals, which are regularly updated to reflect the current status. Counts can be described as datasets containing the numbers of events linked to individuals, populations; or the number of persons, linked to an event, location, etc., in a given time or reference time period. By surveys we mean (usually sample-based) data collections realized by interviews with respondents.

Censuses constitute a special case. They can be regarded as special as special counts, since the data collection is clearly tied to a time period (and repeated usually in ten years). Often they also involve surveys, realized on a whole population instead of certain samples. And, to make a full round, censuses are increasingly based on registers—at least in countries with highly developed administrative data collection systems.

15. A sentiment represented in Stinchcombe's book, for instance (Stinchcombe 2005). This view maintains that the laws of social sciences are essentially the same

as those of natural sciences. Some argue, however, that the former can live on only *ceteris paribus* laws, ones that assert causality *if every other factors stay the same* (Fodor 1974). This is not the place to dig deeper in this issue, however.

16. See Kootstra for a discussion of factor analysis, or Gould apropos of intelligence measurement (Kootstra 2004, Gould 1996).

17. Bulmer is an almost recent example (Bulmer 2001).

18. Not all attires though, I will not address such sophisticated research projects here as, e.g., Falk conducted with his colleagues (Falk et al. 2018). Instead, with my two examples I try to keep thing as simple as possible, while indicating some typical features of these data collections.

19. Loyalty to a mustache brush brand, say, is not an easy case. It is quite a plausible presupposition that many of my friends and colleagues have no mustache, thus the probability of my recommending *should be* relatively low even I am highly devoted to the product. On the other hand, having no mustache does not prevent me or Debbie to recommend the very same brand, e.g., after hearsay, even though, understandably, we have no loyalty toward it whatsoever.

20. An analysis of this bias would be a quite interesting topic in itself, but lies beyond the scope of this study.

21. Observe that the party preference measured in a cohort is slightly different from party preference in the whole population. The difference lies in the very fact that the first is "non-adjusted." The situation resembles contemporary timekeeping with its different time concepts.

Chapter 5

Construction and Truth

One could observe with a bit of oversimplification that everything said so far in this book can be viewed as flirting with one general question: what can be meaningfully said about nature by reading our gadgets? In this last chapter, I will also explicitly address what can be *truly* said about her. By investigating a proposal for a theory of measurement errors, we will find ourselves right in the middle of a dispute on (semantical) realism. It might not come as a surprise that I will defend a constructivist thesis. However, as talking about constructions—without being an architect—it will be immediately needed to set myself apart from some popular movements marching under flags with similar catchphrases. Having set the stage, I will draw my conclusions concerning the validity of our ways of measurement.

5.1 MEASUREMENT ERRORS

The problem of measurement errors is by no means independent of the general problems of empirical confirmation. It is commonplace that errors cannot be avoided; they are immanent in every kind of empirical data. When dealing with errors, one tries to make estimations on the reliability of data, and knowing how apt the data are is inevitable for valid inferences. Not in the least, an appropriate account for errors may be vital for choosing between theories—even if one's taste draws her away from talking about confirmation. In spite of the importance of this question, the literature for measurement errors is relatively poor.[1]

Let us begin with Thomas Kuhn's visual illustration on what he calls the textbook account of measurement (figure 5.1) (Kuhn 1961). I reworked the picture a bit so to fit in better with the recent trends in fashion.[2] The essence

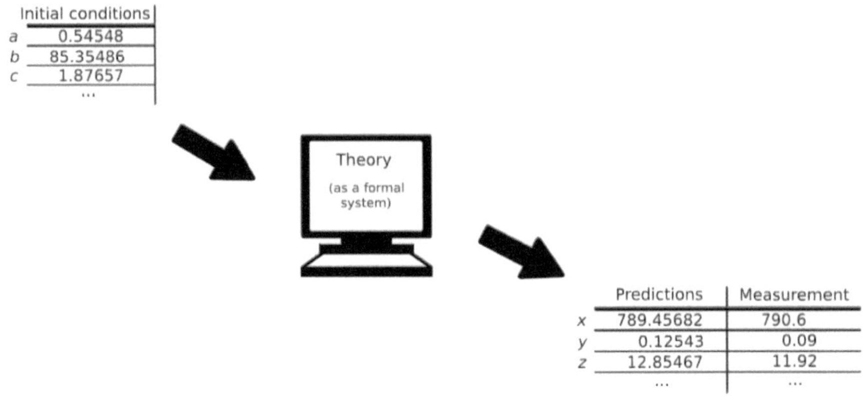

Figure 5.1 Kuhn's Textbook Model of Measurement.

is the same: according to the *textbook* story, the data describing the initial conditions are digested by the theoretical machinery, which produces numerical output in the end. These numbers are then placed against other numbers flowing from empirical procedures (measurement, that is). A reasonable agreement is a positive test for the theory of question. No one is after a perfect agreement, because nature is messy and a theory can never take account of all factors, and because our instruments are imperfect. Indeed, too-perfect agreement is often regarded as a sign of scientific fraud. But according to Kuhn, measurement rarely if ever play the above confirmatory role in *journal* science (which is more reliably reflects what scientists do). Rather, it has major role in the infrequent events called *scientific revolutions* by indicating the anomalies in the prevailing theories. But generally, during the normal course of science, measurement serves as a major tool of "mopping-up," the theoretical consolidation after revolutions.

Though he refrains from dealing with the more mundane functions of measurement (consider, say, that the initial conditions can also be produced by measurement), Kuhn's sentiment seems to echo well in the theories of measurement errors. What I reconstruct here is a theory of Henry Ely Kyburg (Kyburg 1984), but there seem to be a wider consent in the literature on sharply distinguishing between two kinds of errors.[3] *Systematic* errors reflect discrepancies in theory choice; in the face of these errors more fitting language (axioms system) can be found for the empirical data. When measuring distance with a laser beam with nearby objects of immense mass, a theory taking account of gravitational force is more in place. Systematic measurement errors indicate anomalies better than anything else, thus play an immense role in swapping theories, hence in scientific discoveries. By contrast, *random* errors can be seen as a kind of noise in the data stemming mostly from our (necessarily) imperfect measuring methods and instruments, and possibly also from the vagueness of the concepts describing measurables. But how to identify the nature and source of errors, and thus, how to bet on a proper theory is not an easy question to answer, and involves some well-known philosophical issues. Some of them will be touched upon in the next section.

We have already seen how the systems of measurement procedures can be accounted for as formal theories, and how their numerical representations can be established. Sure, when taking account of errors, things get more complicated. According to Kyburg, for every given relation in a theory we should stipulate a distinct, corresponding relation on the empirical data (ibid.). Thus we have:

(a) **R**, a relation instantiated in the models of our axiomatic theory.
(b) **R***, a relation instantiated in the empirical data.
(c) $\mathbf{R}^m$, a relation on a mathematical structure homomorphic to the models of the given axiomatic theory.

The issue gets even more complex when trying to give account of indirect measurement—and most of our measurement procedures are indirect. In this case we must introduce a further relation for the quantity measured indirectly. Systematic measurement, that is, measurement based on scientific laws—the most noble of measurements in Kyburg's hierarchy—requires even more corresponding relations depending on the number of the involved quantities in the theory.

Let us go on with the simplest case. It is clear that **R** and $\mathbf{R}^m$ are linked by their structures being homomorphic. By what links **R*** to **R**? The latter is taken to be a relation imposed by the axioms, which, as such, can never ever be observed directly (not even in principle). The observable relation is **R***. But homomorphism, of course, cannot be guaranteed for their corresponding structures. As an example, take **R** to be equality (in the sense that magnitudes in relation cannot be differentiated by the relevant procedure). Equality is taken to be transitive in most of the measurement systems. But in real, large enough datasets transitivity is often violated. It is a typical feature where errors are to blame.

How to reckon with this situation? As for a less dramatic approach, the authors of *Foundations* try to handle the issue by offering several legitimate ways to live with the crux of intransitive empirical data (Suppes et al. 1989, 300–301). First, and this looks like a really easy road in any event, one might say that the inference leading from observational data to theories is surely an interesting problem, but a practical one, out of the scope of the theorist. Another approach—a bit more hardworking one—is to develop a statistical tool to handle discrepancies in empirical data. One can work out, for example, a theory where parameters are introduced for denoting the relative frequency of x being greater than y in consecutive measurements: $P(x,y)$. These parameters may be well used to "adjust" empirical data to the model. According to a third temper, discrepancies could well serve as nice information sources in many cases: as "errors" may be interpreted as insensibility for nuances. The *Foundations* works in the spirit of all these possibilities: among the neat formal systems for relative frequency parameters and insensibility it largely ignores the untidy staff of practical error handling.

Kyburg, on the other hand, as an advocate of the second approach, also keeps an eye on practice, and tries to give an account of error handling which is theoretically sound but, at the same time, calculates with the arbitrariness of practice. Regarding that he deals with philosophy rather than applied mathematics, he is right: practical issues indeed seem to be an integral part of the story.

Kyburg introduces a language, *L*, for the "ur-metacorpus" of quantitative observational reports (Kyburg 1984, 186–187). The corpus consists of infallible and unrevisable *records*—or protocol sentences in the vocabulary of a long-ago bygone age:[4]

> The fifth measurement of the magnitude x yielded ru;

where by r we mean a real number,[5] and by u we mean a unit for the given quantity. The one above is not the sharpest form of a protocol record, but I will not discuss here (and I do not really have a strong opinion on) how such a sentence should look like.

From this it follows *with zero probability* that:

> The value of magnitude x is ru;

which is bad news. However, goes Kyburg's consolation, we can also infer from the record with a probability close to 1 that:

> The value of magnitude x is between pu and qu;

where the p and q are also reals and r is an element of the interval $[p,q]$; *and* p and q are *chosen aptly*. What is more! It also follows (at least so Kyburg says) that:

> The value of magnitude x is $r'u$;

where r' is real number in the interval of $[p,q]$ different from r.

Clearly, the arguments for **R*** consist of *r*-s and the arguments for **R** consist of *r′*-s; but now the problem is, instead of relating relations, how to have a clue on the relation between *r′* and *r*, at least, in statistical terms. Or, in other words, how to determine the [*p*,*q*] interval. To do this, Kyburg says, we need to accept some principles of statistical, practical, and moral nature (however surprising is the latter).

As an actual dataset is always limited relative to a model of an axiom system, we cannot, for instance, make infinitely many measurements, we need to accept a principle, which we can call the principle of similarity or the principle of statistical inference or the *distribution principle* (ibid.). This guarantees that a sample will always be similar to the whole population in important respects. In particular:

(i) The relative frequency of a given type of error *E* is basically the same in the sample and in the whole population.

This is a rather strong principle, I think, and we have some reasons to distrust it. I am not after the very uncertainty lying at the heart of the whole "inductive statistical inference" issue. Rather, consider the particular and practical problems of measurement in different conditions, or in different orders. Reading equipment in various light conditions, or even before lunch or after lunch, or measuring very little or very large magnitudes may well yield diverse error distributions. It is rather bold to presuppose that our sample will take care of all these departures in distribution. As we have seen above in the case of party preference measurement—for this is a point where social science methodology has something to tell—it is indeed a hard work to exhibit a sample with aligning patterns to a whole. It is not there, it is not readily given, it must be produced with a help of a sophisticated theoretical and instrumental machinery.[6] But all right, let us now go on with the simplified picture.

We need another principle to guarantee that we do not choose the margin of error (interval) too wide. That is why we have to state our practical *minimum rejection principle* (ibid.):

(ii) We should not admit more errors in our data than what we are obliged to admit.

This principle is not what we expect in a methodological context. It relies heavily on our moral convictions—or to be exact, more heavily than the previous principle does. At this point I cannot escape to ask: whence this obligation comes? One might think it must have its origins at a leading authority if along with it someone is bold enough to revise what is readily given by nature. Second, in itself the principle sounds rather vague. Kyburg may also feel this, as he goes on to sharpen it. With regard to measuring length, he states:

> [A]ll the judgments in the set *might* be in error. But to assume this would be gratuitous. So let us take the frequency of error to be the least we *must* assume in order to reconcile the set of judgments with the axioms for an extensive structure. (Emphases original.)

In the context of length, this principle may sound somewhat plausible. Now consider it for an arbitrary structure of magnitudes: when assessing data, we must have a forehand conviction about the underlying measurement theory. Without this we cannot take account of the errors, since our bet on the margins of error depends on how much should we "adjust" the data to be in line with the axioms of the very theory. When we are tired of consequently failing with the adjusting project, then we can drop the theory and start looking for ones with a better fit.

No doubt, Kyburg's account nicely ties a theory of error handling to theory choice, or, if you like, scientific discoveries. And also leaves us with a good deal of questions. For instance: how would I presuppose any theory in a more complex measurement situation? When facing with a large dataset of psychometric research, how would I choose my theory? In these cases, often the very question is, what is our scale like? Or, what if there are many theories within the reach of basically the same effort in data adjustment? How should I choose one?

There is a regrettably popular answer for these questions on offer: once a scientist has a theory in mind, she almost never lets it go. Once she does still, the reasons for theory swap are outside of logic and scientific methodology, in particular, it has social nature. The undeniable fact that theories are always underdetermined by data led many philosophers to relativistic conclusions in different flavors, which, in turn, produced massive reverberations far beyond the realm of philosophy of science.

5.2 CONSTRUCTIONS

While it is commonsense that the account of philosopher, sociologist, or historian on the affairs in science tend to be quite different from that of the scientist himself, many practitioners are quite aware that much more is at play in theory choice than pure observational data. As an example, this is how the physicist James Bjorken put it:

> But whatever the pros and cons for accepting the orthodoxy [i.e., the incumbent set of theories], there are the attendant dangers common to any orthodoxy. With the risk of being banal, I feel compelled to express what I see as the biggest danger, which is that experiments become too sharply focused. While searches for what is predicted by the orthodoxy will proceed, searches for phenomena outside the orthodoxy will suffer. Even more important, marginally significant data which support the orthodoxy will tend to be presented to—and accepted by—the community, while data of comparable or even superior quality which disagrees with the orthodoxy will tend to be suppressed within an experimental group—and even if presented, will not be taken as seriously. (Pickering 1984, 236)

This account leads straight into the contemporary debates on the miseries of scientific practice. What is and what is not under research, which research is rewarded by the academic system and which is not, what are the resulting distortions, and so forth ("Trouble at the Lab" 2013). But let us begin a bit farther out. Hume was the first to raise serious doubts about the principles of inductive reasoning and several philosophers followed him on this road in the twentieth century, as I already indicated in section 3.5 (chapter 3). By the other side of the same coin, Pierre Duhem showed, long before logical positivism got momentum, that logic is not enough for discarding physical theories, because arbitrarily many explanations can be held against the same body of evidence (Duhem 1906). Furthermore, theories are not "alone," but they are always part of a bigger system containing auxiliary *physical* hypotheses, for instance on the functioning of the applied equipment, thus it is never straightforward what to revise in face of recalcitrant empirical data. What he did not seem to intend to say is that we can never have rational reason to choose between theories. This is where Quine took up the issue, already in the fifties but most straightforwardly in the seventies (Quine 1951, Quine 1975). According to him, theories are so seriously underdetermined that one can hold any theory in the face of any evidence, and one can revise *any* (not just the physical) part of the body of knowledge when needed.

At this point relativistic ideas broke loose. If there are no rational ways to choose between theories, each of them is equally valid, science cannot provide ways to tell which theory is better. Temporary prominence of theories is the product of contingent social and historical proceedings and events. As for one well-known example, Thomas Kuhn (Kuhn 1962) argued so. He attacked the view that old theories are often limiting cases for their successors, in the sense that explanations and predictions of the old theory are a subset those of the new one. In other words, he denied the cumulative nature of scientific knowledge. He observed that in any given period of what he calls normal science, the machinery of scientific research is driven by incumbent *paradigms*, which regulate the whole practice from valid research problems through criteria for solutions to acceptable experiment protocols. Paradigms are considerably resistant to *anomalies*, but all of a sudden they may change in revolutionary ways. Now the resulting new paradigm is *incommensurable* with the previous one. There is no way for rational comparison, since all of concepts and methodology are strictly tied to a given paradigm frame.

In another famous book of which several updated editions were issued over the decades, Paul Feyerabend devotes himself to the task of "liberating" scientific method (Feyerabend 2010 [1975]). He rightfully emphasizes that science cannot and should not be built on methodological rules that are universal and fixed for good. From this he reaches the dubious anarchistic conclusion: "anything goes"—meaning that anything goes as scientific methodology. If this is taken to mean that there are no golden rules, one always have to rely on her intuition, creativity, and insights, we may even live with it. But as we read it as instead of accelerating, colliding, and measuring particles, it is equally fine to turn to an oracle, it is simply untenable. Unfortunately, Feyerabend provides reason for this latter reading, as he explicitly compares scientific knowledge to myths. It is to be noted, however, that interpreting Feyerabend is not easy, as he deliberately and admittedly plays around and brings his conclusions to the extreme, just to distance himself from them at other places.

Then came the movement of sociology of scientific knowledge in the works of, for example, David Bloor, Barry Barnes, Henry Collins, and Bruno Latour. They took the task of investigating the social, cultural, institutional, economical aspects of science upon themselves—which I would deem a nice and apt choice. Meanwhile, though, they have been spreading the word of relativism, sometimes explicitly and proudly attaching the tag on themselves. Many arrived to the conclusion that world or "Nature" (Latour 1987) has no role in scientific knowledge at all, it "in no way constrains what is believed to be" (Collins 1981). To put it roughly, if there are no constrains of nature and anything goes, every thought on the world is equally legitimate, they are only classified by contingent social conditions, accidental cultural trends, and incumbent political power. These sentiments were predestined to gain ground quickly outside the esoteric circles of philosophy of science. The thought the "everything boils down to subjective perspective and interests" (Laudan, 1990b, X) provided to be fertile in the post-modernist, post-structuralist milieu of the era, and led, on the one hand, into "science wars" waged first between academic circles of "hard" and "soft" sciences (Sokal and Bricmont 1997). Nowadays these wars are also fought fiercely within social sciences themselves (Pluckrose, Lindsay, and Boghossian 2018). Outside the academic world, the battle rages on around the "post-truth" nature of common talk with far-reaching consequences on the politics in Western societies.[7] Being around now for several decades without any sign of a decline in its vitality, relativism is a stubborn symptom.

This is, of course, not the place to discuss these developments in detail. This brief syllabus serves only to provide a context for my explaining myself about using the words: constructive, construction, constructivism. For in line with the developments sketched above, constructions are often meant *nothing more or less than* as social construction. Ian Hacking wrote a book with the title, *The Social Construction of What?* (Hacking 1999). In it, he writes: "talk of social construction has become common coin, valuable for political activists and familiar to anyone who comes across current debates about race, gender, culture, or science" (ibid., 2).[8]

As the title of his book and the volume of the processed material suggests, Hacking is a real expert of social constructions. It is therefore an apt choice to try my constructivist views in the face of his theses. Of course, I will restrict myself to the context of science. Science as a human endeavor is necessarily part of the social realm. Its constructions, such as its concepts, theories, institutions, and gadgets, do not go without social aspects. Who would deny that my most mundane construction of an apple pie contains social ingredients, anyway? But apples are not social, not even if their productions requires human activity. Nor is the fact socially construed that I can survive with the pie instead of deceasing in hunger. Hacking also draws to similar conclusion by insisting on a distinction between on the one hand science as assemblage of truths, which is not social, and on the other hand science as activity, which is trivially social (ibid., 76). Unfortunately, he does not let us know how truth to is be understood. It would be all the more interesting to know, because he maintains a dedicated purgatory for the suspicious "elevator words," such as fact, knowledge, reality, and truth.

Hacking identifies three main areas where the realist and the constructionist are necessarily in disagreement, the "sticking points." I am a bit puzzled whether it is meant as a dichotomy. Of course many philosophers are not concerned with what is going on in science at all. As I may rely on Dummett's analysis of realism (Dummett 1993), certainly I am not a realist concerning the statements of science—nor concerning the statements *about* science, as it matters. But I cannot fully buy Hacking's constructionist points either, not the least because they are not always crystal clear, much depends on their reading. Maybe, in Hacking's world I am just a "constructivist" not a "constructionist" if it makes any sense at all.

Now for the sticking points. Hacking's first point is contingency: the constructionist holds that no scientific proceeding is necessary, there can always be alternative, equally successful programs. The second brings in the old question of nominalism as a problem of structures: there are no inherent structures of nature, they are to be established—so says the constructionist. Finally, the realist and the constructionist differ in their explanations of stability, the latter holds that canonical scientific beliefs always has social elements. I will look at these points one by one, in a reversed order, finishing with the first point, which I find the most important.

Hacking summarizes the view on stability as this:

> The constructionist holds that explanations for the stability of scientific belief involve, at least in part, elements that are external to the professed content of the science. These elements typically include social factors, interests, networks, or however they be described. Opponents hold that whatever be the context of discovery, the explanation of stability is internal to the science itself. (Hacking 1999, 92)

Internal of what science? The science as observed or the science of observing science? Physics does not make statements of the stability of its theories (though *physicists*, of course, may). No such concept like endurance of evolution theory exists in the vocabulary of biology. If sociology makes such statements, they are probably and unsurprisingly of social nature. And the stability of social statements is also explained by social factors, as the reflexivity principle of the strong program of sociology of scientific knowledge requires (Bloor 1991 [1976], 7). But normally, with probably further exceptions as meta-philosophy, reflective statements on scientific activity is not part of the corpus of the discipline in question. Surly not in natural sciences which are usually meant above all in similar discourses.

But all right, let us take up the challenge with trying to give a plausible interpretation to this point. For instance:

> Someone, who looks at some tenacious scientific status quo and finds that it has to do only with the rightfulness of the given class of scientific statements, is a realist.

In this case, there are not so many realists among the ones who theorize over science. For hardly anyone denies that conventions, that is, petrified social agreements play a crucial role in the success of a discipline. Without them, we could not even make and understand statements. Okay, then let us put it this way:

> Someone, who looks at some tenacious scientific status quo and finds that apart from conventions it has to do only with the rightfulness of the given class of scientific statements, is a realist.

Now consider someone who holds that had the library of Alexandria not burned down in the first century BC, we would have considerably different picture on the economy of ancient times. The burning down of the library is not part of the history of economics, nor is regarded as a statement of it, but as an accidental event. Is the given economist-historian necessarily a constructionist, then? I am not convinced. Not the least, she may well hold that each statement of the history of economics is true or false, independently of the fact that we do or do not know which. Probably Hacking has slightly different senses of realism in mind. It is common to suppose that the realist holds things like theoretical concepts refer and the successful theories are approximately true. I cannot see it makes much difference. In any event, though I cannot see it as a real sticking point, I willingly admit that social factors play role in the hegemony of certain scientific discourses.

More interesting (or more sticky) is Hacking's second point. Are the structures "out there" or are they added by the scientist? I do not "dislike distinctions" (Hacking 1999, 67), so I would set two kinds of structures apart. There are structures which are entirely human-generated, and there are those which are reflected and coded by human activity. Note that none of these kinds are devoid of social elements.

The first type of structures consists of conventional systems and formal games, such as the Latin alphabet and graph theory. One could object that the elements of the alphabet represent dedicated phoneme types, but this is not exactly the case. Orthography has to be learned, in other words, the conventional link has to be established. For instance, the symbol "k" stands for quite different (types of) sounds of speech even within one language, such as English or Swedish. Consider the words *tick*, *knight*, *kirke*. True, alphabet is *used* for recording speech, but the interrelation is conventional and sloppy.[9] As for my other example, it is not my intention to dive into long argumentation on the foundations and nature of mathematics. It enough here to state that mathematical "objects" are in no way in the reach of our empirical means.

Now for the second type of structures. Different kinds of thermometers merged in the same vessel of water on the stove provide similar enough behavior to state that they reflect a structure in nature. Individual trees in a lane shows an order which can be reflected by some of our mathematical means. It is to be made clear, however, that no structure in nature is directly accessible for us. Whenever we are to exhibit a structure, we need a rather more than less rich machinery of the first type of structures. For instance, both of the above mentioned structures, alphabet and graph theory, are of great help when we are to make statements about neural network topology. Without them we could not create any piece of codified scientific knowledge.

As for the context of measurement, I noted earlier that I cannot side with Stevens, who, rather openly than tacitly, held that we can stipulate scale types opportunistically and according to our research interest whenever needed. A common perception of measurement is that when measuring we relate empirical data structures to mathematical ones. The first is not entirely human made, even if we use conventional and formal structures to represent them, and even if nature rarely tells us strictly what exactly to use. But the answers of nature may guide us in our choices, so, we can establish congruence by iterative tests. Were it not the case, the whole project of measurement theory (and even measuring!) would have little meaning.

Finally, back to my two distinct types—it is to be noted that there may even be borderline cases of structures not fitting well in either of the above classes. For instance, I would be reluctant to decide which of the two types certain *taxonomies* belong to. But my point is straightforward still: human activity is essential in establishing structures. It does not mean at the same time that they are entirely arbitrary. Nature has a word in them.

This leads us straight to the first sticking point, which has to say a lot about arbitrariness: at each phase of history for every achievement there can be an alternative, equally successful research program. Note the term *research program*, invoking Lakatos (Lakatos 1970). Indeed, Hacking explains the notion of success with his insights (Hacking 1999, 70). A research program is empirically and conceptually *progressive* if its new theories make new predictions a richer but simplifying conceptual armor, while keeping the most robust predictions of the previous theories intact. They are *degenerative* otherwise. Hacking does not let us know whether progressiveness equals success or if it is only the most important part of it, but in any case we can take it as a differentia specifica.

To stress the point of the constructionist, he turns, with overt sympathy, to Andrew Pickering, who arrived to his own incommensurability thesis while historically analyzing a period in high-energy physics in his book (Pickering 1984). Particle physics with its quarks as we know it today is a product of contingent social, historical, instrumental factors. It was not necessary at all for the quark model to rule, the development of physics could have followed a non-quarky way, "old physics" could have provided to be equally progressive and led to the same success. It did not, and this fact steals a bit from the strength of the argument. Anyhow, according to Pickering, the two stages of particle physics is incommensurable. It does not mean that they are logically incompatible, rather that their conceptual, phenomenological,[10] institutional, and instrumental embedding are so different that they speak entirely different language. There are no ways whatsoever to compare them, they live in quite different *traditions*.

That the course of history could have been different is not an overly stunning thought, and, as I already argued in the case of stability, it is perfectly compatible with realism. The contingency of the present state of knowledge does not confront with the sentiment that every well-formed statement is true or false, independently of our knowing which. True, Pickering professedly does not care about truth. For me it seems rather that he is playing some *yes*, *no*, *black*, *white* game when trying to avoid the words *true* and *false* (especially in his *The Mangle of Practice* [Pickering 1995])—in which game he is rather clever, it is surely to be admitted.

But back to comparing research programs. Are there actual examples for equally successful, incommensurable programs? If there are, how could we know at all that they are about the same "object," and we are not just trying to compare apples and oranges? For progressive programs in ethnography, ornithology, and linguistics are incommensurable for sure (even if one may find their common origins in ancient philosophy). So what ties the incommensurable comparables together? A ready answer to this that they rely on the same set of data. But relying on the same bundle of tapes recording interviews with forest-living natives, ethnography, linguistics, and ornithology may reach entirely different conclusions verifying or falsifying their actual theories. If this is what is meant by incommensurability, I cannot see any sticking points around. But all right, let us be more serious. As the relativist claims that all the data, their interpretations (not to say institutions equipment and stuff) are all products of arbitrary social constellations or traditions, how do we know what to compare to what?[11] In particular, have any data produced after a tradition split anything to do with the old tradition?[12]

In zooming to my view, let me take an even more simplifying jump. We began with particle physics, let us arrive now to glasses and cups. Glasses and cups can be found all over the world, embedded in different traditions and reflect quite wide range of cultural and anthropological factors. Could they be more different? They are different in volume, shape, and ornaments on a wide range, they have different names, they are involved in different rituals. Still: are not they similar enough? For instance, under certain circumstances (they are held properly, they have no holes, etc.) all are able to hold back liquid.

I do not want to appear ignorant to Pickering's great work in *Constructing Quarks* (Pickering 1984). I appreciate his thorough historical analysis and his trying to show that the "quarky way" was not necessary, that particle physics could have developed in a different way. But even after swallowing the counterfactual, and presuming clear and definite boundaries for a discipline, the point comes down to the realization that theories and their conceptual machinery are in many respects accidental. To invoke the aviatic example as obligatory in this context: I can imagine a stage of aerodynamics without the concept of turbulence where airplanes still fly.

In his later book, *The Mangle of Practice*, which is otherwise full of really amusing case studies, Pickering offers a picture, or rather a metaphor on the role of nature in scientific progress (Pickering 1995). He describe scientific activity as a *dance of agencies*. Human agency—the scientist—sets up her equipment and tests it against material agency. The human agency becomes passive at this point and just reads the gauges. Now material agency has a turn, it becomes *temporarily emergent* and usually the test does not work out for some reason. Nature resists. Then the human agency accommodates the resistance, modifies the equipment and tests again, and so on. This "dialectic of resistance and accommodation" (ibid., 22) is what "mangles" the content of science until a stable phase, a *robust fit*.

Note that according to the view the emergence of material agency is something like blind chance, nothing is determined beforehand. Hacking writes:

> The constructionist about (the idea of) quarks thus claims that the upshot of the process of accommodation and resistance is not fully predetermined. Laboratory work requires that we get a robust fit between apparatus, beliefs about the apparatus, interpretations and analyses of data, and theories. Before a robust fit has been achieved, it is not determined what that fit will be. Not determined by how the world is, not determined by technology now in existence, not determined by the social practices of scientists, not determined by interests or networks, not determined by genius, not determined by anything. (Hacking 1999, 73)

I do not have strong opinion on determination. But if it was really all that the "upshot of the process" is not entirely "predetermined," I would willingly agree. Rather it seems to me that Pickering and Hacking are playing around with offering something and then trying to get it back with the other hand. First they admit that there is a material part in the story on science—we can ask nature and she answers. Then they hastily add that it is not an occasion for celebration. Because nature's answers are entirely random, and only temporarily emergent. It does not make much sense for me. Does it mean that reproducible experiments and airplanes flying from time to time are no more than a random (however robustly random) fit in every case? What else robustness could mean that—yes, by the help of traditions, institutions, machines, and all—a certain *regularity* was established?

As stated above, with his sticking points Hacking seems to suggest a dichotomy, as if the only way to realism would be the denial of all social aspects of science, and the only way for an antirealist would be to accept some kind of constructionism. I hope to have shown in this section that it will not do. As for me, I do not hold that every statement of science is true or false independently of our knowledge. Nor do I deny the social nature of science. What I deny is that it is exclusively social.

Consider a rake. We could never stop stressing how much a rake is a social construction. Its concept and name(s) are socially constructed. Its fabrication and use are culturally embedded. And so on. Right. It is still quite painful when that construction smashes to the forehead after an unwary step while sneaking home through the garden at night. Did nature resist? I even venture that the lump on the head is not socially construed at all.[13]

5.3 MEASUREMENT AND TRUTH

I am convinced that the problem of truth should be addressed instead of putting it on a shelf like Hacking, or overlooking it with a blank face as Pickering does. I will not take up the ambitious task of building a meaning theory in this short section, of course. What I do here is just provide a semantical analysis of a narrow class of statements: the statements of measurement results. Certainly, it may have morals for a wider field.

With what already has been said on procedures, structures, errors, and realism in mind, let us take again the banal case of length. Having a rod, which we can also honor by the name *r*, we can illustrate the realist stance with the following statement:

(*Real*) The sentence "The length of the rod *r* is *x*" is true or false; independently of our knowing which (and of our inability to know it ever).[14]

It is now easy to see what is the most unfortunate feature of a consequent realism. As we have found, the realist presupposes that (at least some of the) quantities are continuous. Again, we have realized that every measurement is error-prone and can be deployed within certain margins of error only. And, what is more, we realized that we cannot ever arrive at an irrational part of a unit by measurement—not even in principle. In particular, this means that we can always consistently replace whatever body of measurement data by the elements of the set of rationals. We must, then, see that every sentence asserting a measurement result is, to very high level of probability, that is, practically, false.

By contrast, the operationalist—an outright antirealist—denies (*Real*), but holds dubious views on meaning and truth at the same time. For if we maintain that measurement is mere assignment of numbers to objects—as the Stevensian does—we must conclude that every sentence asserting a measurement result is trivially true. Since measurement is only assignment, we measured a magnitude once we assigned a number according to an arbitrary rule. Even by the more modest, instrumentalist, Bridgmanian version of operationalism, the very concept of a quantity is brought about by the given procedure we use. Hence, a given magnitude is constituted by a given measuring act, and nothing more.

In short, a statement like:

The length of rod *r* is 123.5 cm.

is *always false* for a realist and *always true* for the operationalist. What is more, the realist does not even require a measurement procedure involved for this, apart from a conventional agreement on units. By contrast, the operationalist—either an instrumentalist or an advocate of mere assignment—requires a measurement procedure for such a statement to be assertable. A further question what else he requires. Anyhow, both views are at odds with our intuitive or common sense perception of measurement.

What a realist would answer to this is the following. Sentences stating results for successful measuring acts can be regarded as approximately true. By this we mean that we can always assign an interval to a measurement result which covers the *real* value of a magnitude with extremely high probability. Practically that means that every sentence asserting a measurement result should be understood as if a "plus or minus x" would be amended to it. So far so good, but now the realist has to face with the problem of determining x for an appropriately high probability. That is, she has now the not so trivial task of translating every sentence of measurement to the "real" language of intervals.

The operationalist could object as follows. I, Bridgman, can easily lie about my measurement results to my assistants. In fact, I am a bit afraid that my assistants would (let it, inadvertently) lie about their measurement results. So a statement like the one above is not always true. And truth is a private experience anyway. Well, okay, but at this point, where measurement cannot legitimately made public, we can doubt that we are by any tiny a bit better off. Again, can Stevens say the same? No, as long as he is consequent. Measurement is nothing more than numerical assignment. But what else is uttering the above sentence if not numerical assignment?[15]

Still, the modest (instrumentalist) operationalist approach and also the realist one have some morals to consider. First, there is no sense in talking about quantities and magnitudes without having an effective measurement procedure at hand. Second, the deviations in our actual measurement of a given magnitude often tend to fall into a given interval. As with technological progress the precision of our equipment is increasing so gets this interval narrower, but, of course, never closes to zero, always remains an interval.

With this picture in mind we can endorse a *constructive* approach: the continuum of a quantity can be conceived—entirely not independently of the appropriate measurement procedure—as an ever-changing chain of overlapping (fuzzy) intervals. Thus it is somewhat *natural* to search for homomorphisms for our measurement systems in the realm of intuitionist reals (or a substructure of them).

In a constructive setup, when establishing equality between two magnitudes by a measuring procedure, we can say they are equal within our current margins, that is, so to say, the initial segments of their "approaching" free choice sequences are identical—and that might not always be the case. What we cannot do: we cannot say before any measurement that one magnitude is either less, either equal, either more that the other. We can say only that they are in a relationship analogue to one of the intuitionist reals. Two of the infinitely proceeding sequences (as two elements of the same spread) are equal if their *n*th components are equal for every *n* (Heyting 1956, 36). Practically that means that when exhibiting the consecutive elements of two series, we find them equal as all of their elements are equal until *n*. Should we find that they differ in their $(n + 1)$ th element, we would not regard them as equal from the very moment we exhibited those elements. Hence, when failing to meet the requirement of the transitivity of equal measures, so that $a \approx b$ and $b \approx c$ but $a \succ c$, we do not have to give up our belief of measureability.[16] For we just compared the *n*th element of our sequence of measuring *a* and *b*, the $(n + 1)$th of our measuring *b* and *c* and the $(n + 2)$th of our measuring *a* and *c*. The transitivity of equality is not necessarily violated. But it is when these sequences entirely depart from each other.

Now, how about errors? On might justly feel that there is no real room for an error concept in a constructive approach. But if we take a look at the discouraging complications that burden an error theory, we can deem it a virtue rather than a loss. Still, we cannot just stipulate sequences, we need some empirical anchors. Let us take again the notion of truth. Of course, the constructivist cannot hold (*Real*), in fact, explicitly denies it. Instead, he holds the following:

(*Con*) The sentence "The length of the rod *r* is *x*" is true or false in virtue of our valid measurement procedures.

The constructivist has to answer the question: how to assess the validity of our procedures? But first note how (*Con*) sets constructivism apart from operationalism. If the constructivist held that the meaning of quantity terms and the respective sentences are fixed by their measurement procedure, he would also hold by (*Con*) that the meaning of sentences is determined by their truth conditions. But this one he denies, as he maintains that truth and falsity can only be asserted in virtue of some knowledge, proof, procedure, or construction, and, at a given point, entirely meaningful sentences may be devoid of truth-value, possibly for good and all. Consider these examples:

(i) Every even integer can be expressed as a sum of two primes.
(ii) The number of the elementary mercury atoms enclosed in my favorite thermometer is even.
(iii) The radioisotope thermoelectric generators of Voyager 2 generate less than 260 watts at the moment.

As for (i), the fact that no one has ever found an exception does not make the statement true. For it to be true, a proof would be needed which meets all the requirements placed against a proof by the community of mathematicians, and which is accepted by most of the experts. By contrast, for it to be false it would be enough if someone could exhibit an even number of which it can be shown that it cannot be reached by the addition of exactly two primes. As none of these conditions are fulfilled at this stage in the history of mathematics, (i) is neither true nor false.

As long as our atom model goes, and as long we exclude evaporation, ongoing chemical reactions, and other means of escape, (ii) is quite clear-cut. The number of atoms is countable and finite. Nevertheless, I am not aware of any viable means of counting atoms in significant amounts of material. Even if such exists, its application for a closed system would probably be more than troublesome. Fortunately, (ii) lacks any scientific interest. Still, the statement is meaningful, and we do not have any grounds to decide on its truth-value.

There may be more at stake around the truth of (iii). I am not sure whether Voyager 2 actually sends diagnostic information on the power of its generators or not. (I would not think it does.) But only suppose we can send a request for this information at t_1, and Voyager 2 sends its fine-grain measurements "immediately" upon receiving our message. It would take more than 33 hours (NASA Jet Propulsion Laboratory 2018) to get the result at t_2 (at the place we associate the clock readings with). Now, how to understand "at the moment?" One could argue that we *in principle* have no way to know what is the situation "now." Someone else could (rightly) insist that we can cope with simultaneity in an operational way: all there is to be known is that the remote event happened (conventionally) half-time between t_1 and t_2. On the other hand, we have ample information on the buildup of the probe, the parameters of its supply system, and the half-life of the isotope applied in its generators. Based on this information we can make some calculations and estimate the sum power of the generators "around" t_1. Although we may deem it a less fine-grained result, still we can conclude that in virtue of our more or less reliable information (iii) is true.[17]

So again, as shown, the constructivist cannot hold that statements are just true or false by themselves, nor that their meaning lies in codified measurement methods. As for meaning, here we only tentatively suppose that the meaning of terms is given by how they are used by competent (expert) speakers and the meaning of sentences is given—roughly and typically—by the meaning of terms and their way of composition. As for truth, we need some more work to show the place of empirical factors, at least in the small realm of evaluating statements of measurement results.

The above examples suggest that truth heavily depends on institutional and instrumental factors. This is so. A possible proof for (i) should be canonized by the community of mathematicians, the truth (or falsity) of (ii) could only be established with an appropriate tool at hand. Measurement in cases like (iii) relies heavily on the actually existing theoretical background. From this, one could conclude that truth or falsity depends entirely on contingent social and historical factors. And this is not so. I can admit that much that measurement procedures has to be canonized in order to make them apt for determining which statement is true, which is false. Neither my measurement of length by my steps, nor Bridgman's subjective experiences provide enough grounds for a valid measurement procedure. At least in the context of science. The validity of a procedure is context-dependent, and the context of science and engineering requires reliability (reproducibility) and publicity in measurement.

Again, a bet on the validity of a procedure may contain complex theoretical considerations, but also pre-theoretical insights and anti-theoretical commitments. These latter are our main concern here. Let us take again Regnault and his thermometers. When he identified a group of gases showing similar enough behavior under the same circumstances (i.e., expanded on the same or similar enough scale while heating), he bet on this group as the best materials for a thermometer. He did not supposed it to be "right" kind of thermometer, only that here we exhibited congruence (nature affirmed the approach), and this fact places the identified materials above the ones where we observed diverse behavior. Regnault's approach can be said to be anti-theoretical in the sense that confessedly and deliberately avoided indulging in speculations around the reasons of the behavior of different materials and any law-like relations between physical quantities. Regnault meant to rely on his experimental aptitude only.

Still, Regnault's approach did not go without—and actually no experiment can go without—some minimal theories, or tacit presuppositions. For one, he surely supposed he had got in possession of *different* gases from time to time, and this conviction could hardly be reached without theoretical elements. He possibly had theoretical grounds in believing that his vessels do not bring about some false regularity, as if some wizard would distort them to behave so. He also possibly surmised his own sharp-sight. And a good deal more. Of course, what he would have admitted to be theoretical and what he just tacitly relied on, we cannot tell. But even with his presuppositions, his accomplishments work as a fundamental finding on the ways of nature. The stage he arrived at shows considerable stability (as Hacking or Pickering would put, it reached a robust fit), his experiments are reproducible, his findings eventually provided to be apt for an embedding into a subsequent theoretical framework. But above all, his results provided grounds for a consistently applicable temperature scale *independently* of the latter theoretical support.

In general, the exhibition of congruence may involve much more sophisticated theoretical machinery and even delicate means of institutional coordination as we have seen in the case of timekeeping.[18] But it should not necessarily work this way, and epistemic iteration always has some anti-theoretical feats. The whole story of theorizing has to begin somewhere. For instance, when we establish some congruence by our iterative procedure, we do not and should not presuppose that there is a magnitude ascribed to some object of study M_r staying constant between t_1 and t_2. Not in the least, at first we do not have theories on simultaneity or time spans. (And generally, it is also rather doubtful if we should presuppose such thing in any case by default.)

In an ideal neverland, only after reaching a significant body of sensory data showing regularities in possibly several respects do we arrive to a point where we build our laws and theories. But we can identify quantities either by sophisticated theoretical machinery, or by naive insights, and then we can begin to deal with calibration. That is, we can begin to sharpen and canonize our procedures for measuring magnitudes of a given quantity with the help of congruent phenomena.

"This convergence provides a basis for a workable notion of accuracy. We can say that we have an accurate method of measurement, if we have good convergence," Chang writes (Chang 2004, 217). I think we should be more cautious in talking about accuracy. I would rather prefer terms like reliability or resolution. But apart from this, he is right. As values, which interestingly (and misleadingly) can be both called "real" or "ideal" in this context can only be seen as blurred and non-accessible limits for iterative operations, by reliability or resolution we can mean some intuitive (even if technical) assessment of procedural aptitude. Chang says: "truth is a destination that is only created by the approach itself" (ibid., 217). No, truth is not a destination, but an intermediate station, though it is indeed created by, or better, can be ascribed in virtue of the approach. Iterative, canonical procedures by which we may track congruent phenomena provide our only grounds for the assessment of truth or falsity of our statements of measurement. This way they are really important constituents of knowledge—may whatever be meant by this last concept. What is to be admitted anyhow: truth is neither future- nor past-proof.

A final note. To say that none of our individual pieces of measurement can be in error does not mean that the constructivist cannot have expectations on the outcomes or theoretical considerations for classifying the results. She may well have. Consider a circle drawn in the sand of the beach. The law on the ration of diameter and circumference will likewise be guide for the realist and the constructivist when making measurement on the properties of this circle. This ratio may provide grounds for explaining correlations. Now the problem for the realist will be to account for the lack of correlation! For he surely insists that all of the measurements are in error, but he probably will be reluctant to give estimations on the error margins, despite the laws and theories applied in the procedure. Perhaps he would be tempted to say that this circle cannot even be measured *properly*.

5.4 CONCLUDING REMARKS

In one form or another, most of the problems raised in this book have been discussed many times during the course of history in analytic tradition. These issues, however attired, escaped permanent settlement from time to time, just as the concerning general rule for philosophical problems requires. Here, of course, I cannot hope to have solved all (or even any) of them for good and all. Still I hope I could (constructively) contribute to the ongoing debates in the philosophy or epistemology of science with my own insights. In particular, I did not dare to take a task like building a general meaning theory for scientific statements.[19] My more modest aim was just to show a plausible semantic interpretation for an important subclass of them. And I do hope I succeeded in this humble task.

My train of thought went as follows. I observed that the conceptual armory of the early representationalists is flawed when it comes to measurement. For one, the status and aptitude of their rules are more than unclear. Again, concatenation, an empirical procedure type central in their theories, is unfeasible as a basis for a consistent, non-circular concept of quantity. In the light of all these, their classification of quantities also fails. The investigations lead to the conclusion that our equally conventional and operational ways for exhibiting congruent phenomena via empirical procedures could possibly serve with the needed grounds.

I analyzed some awkward peculiarities of a realist stance with regard to quantities and continuum. I showed that a bet on an "objective" continuum is not only metaphysically burdening, but also gratuitous. A fragment of the rational numbers are enough for representing any kind of empirical data as there always exists a consistent renaming. After taking notice of the vital need for continuous functions in sciences, I proposed to endorse the concept of real numbers and continuum in constructive mathematics (along with an intuitionistic logic).

After surveying some of the main points of the everlasting discussions and controversies on social science measurement, I arrived to the conclusion that—in addition to the datasets trivially measurable by counting—relying on congruent phenomena is, to an extent, open for these disciplines as well. One example was the application of conjoint measurement, and I indicated that there may also be other ways. In a minor case study on the dubious practice in some ventures in preference measurement I implied the in-principle possibilities for improvement also.

Turning back to issues of realism and constructivism I tried to make clear what is and what is not implied by my constructivist approach in the light of a general constructionist sentiment. I cannot say I succeed. Constructionists incriminate some suspicious words, then they either still use them or replace them with likewise suspicious words or whole metaphors. Hence, it is not entirely clear what is implied by their stances. Anyhow, the least we can say that one can be an antirealist without embracing any radical social constructionist view. In particular, one can deny that every well-formed statement is true or false independently of our exhibiting which, without holding that nothing is non-social.

I hoped to show that both the consequent realist and the pure operationalist must hold an awkward position in the face of the simplest sentence stating a measurement result. Namely, it is always false for the former and always true for the latter. A way out may be to embrace a constructivist view, where valid procedures indeed grounds truth or falsity, but do not determine meaning (though may add to it). The assessment of validity is in most cases not a straightforward task, though, and here communities and institutions indeed enter the picture. Exhibiting congruence by a proper instrumentation may constitute procedures in virtue of which we may legitimately assess the truth values of relevant statements. There is surely a spectacular loss (if it is): we cannot name any errors around.

I would not deny that the results presented here leave several problems open. Here I just mention two major ones, one for methodology, one for semantics. They may also mark the way forward for further investigations.

As for the first, it is entirely not clear at this point how the search for procedures to exhibit congruent phenomena could be worded as a general methodological principle. These procedures come in many forms, from applying conjoint structures to comparing atomic clocks. Sometimes they require only a minimal set of presuppositions plus experimental aptitude, in other cases they rely on deep theoretical background. It happens that they cannot go without a rich theoretical, instrumental, and institutional machinery, plus a large bag of ad hoc commitments. At the first approach, it is hard to see the common grounds. Probably the only principle mentioned in this study is epistemic iteration. Here we are likewise in need of further analysis, but at least it indeed sounds like a good candidate for the main concept in a principle for scientific research. It is to be seen, however, that on normal Sundays we are rather ambiguous with iteration. Consider a mother measuring the temperature of her child. She often repeats her measurement if it does not meet her expectations. Never if it does.

Now for semantics. We played a bit around *meaningfulness* in this text, that is, around the criterion of sensible implications from measurement procedures. As we have seen, keeping an eye on invariant properties may help us to cogent inferences. It is still unanswered by my inquiries how the *meaning* of a concept is constituted, and what role truth plays in its constitution. Chang proposes that by conformity and overlap of procedures we can guarantee meaningful concept broadening or drift. But this possibility is rarely if ever available in practice. Concepts usually tend to behave in much more untamed and naughty ways. In a typical case it is really hard to tell what measurement, as a valid procedure for exhibiting truth, adds to their meaning.

Let us take a case which we can call the Norwegian shoreline problem. The concept of the length of Norwegian shoreline can be seen to be hopelessly vague, but if we are a bit touchy on vogue, we rather call this length a fractal. Fractals are mathematical objects with fractional dimensions, usually arrived at by some endless iterative operation. In the case of the length in question, the result of our measurement is highly sensitive to the density of the points between which distance is measured, that is, to the resolution of our data collection. The more point we have to work with so the shoreline is getting longer.[20] One could arrive to the conclusion that this length concept has no meaning at all, unless we suppose that country limits may be of fractional dimension, which, I think, neither Norwegians fishermen nor border patrols do. As for me, I think the length of the Norwegian shoreline is a meaningful concept. I also think that we can make true statements on it—even though it remains a question as to what is the relation of truth and meaning in a non-truth-conditional semantics. For instance, it is not of fractional dimensions, since it is not a mathematical object. It is also not pink. What is more! We can insist that the length of the Norwegian shoreline is *n* meters, in virtue of our recent and canonical measurement procedure. Nevertheless, is not clear how exactly the original concept is shaped by this.

NOTES

1. Interestingly, the authors make a false reference for a probably never written chapter of *Foundations* on errors (Krantz et al. 1971, 13). However, the opus does account for "errors"—as threshold representations (Suppes et al. 1989, 299–383). Again, Campbell also proposed an error theory somewhat similar to the one discussed in this chapter (Campbell 1920). In the contemporary literature the issue is traded as the accuracy problem (Tal 2011, Teller 2013, Grégis 2019).

2. The original image in Kuhn's paper features a hand mill (Kuhn 1961, 163)

3. See Berka as an example (Berka 1983, 196–198).

4. Interestingly, Kyburg denies the possibility of such infalligible sentences at another place (Kyburg 1984, 234)

5. We rarely use others than rationals in measurement reports, but we should by no means exclude this possibility. We may easily let someone report a result as. (Even though, as I have proven in section 2.5, rationals are *enough* for reporting measurement reports.) Well, she may even use hyper-complex numbers, but let us keep things simple.

6. It is not entirely clear how to interpret the principle. Either it says we should always suppose the same error distribution in the sample than in the population, or that we must take care of such a distribution by all of our means.

7. If Laudan called American political campaigns the most "pernicious manifestation of antiintellectualism" at the dawn of the nineties (Laudan, 1990b, X), I really wonder how he would adjust his words to the recent situation.

8. Most probably, the term became in fashion after Berger and Luckmann (Berger and Luckmann 1966).

9. Forget now the linguistic theories on this feature, I do not think they are relevant in this context.

10. Phenomenology in this context means the first-line interpretation of accelerator data.

11. Not in the least, whence constructionists get the strength to peek out from one tradition and talk about incommensurability? Reference, truth, and other stuff are not on their agenda, still it is worth a question: what do they do when typing a sentence on social constructions on their socially constructed keyboard?

12. Let it that they have. Even in this case we can keep with Laudan's realist: "You claim that because no one has shown that resurrection of Aristotle's biology [claiming that offspring entirely derive their traits from the father] is impossible, we should regard 'the weight of reason' . . . as equally balanced between that paradigm and its successors. I think that is crazy. . . . [It is] conceivable that Aristotle's biology might be successfully revived. . . . But that is *not* the same thing as saying that, given current evidence, it's as reasonable to believe that human beings reproduce in an Aristotelian fashion as it is to believe that they reproduce in the manner of modern embryology" (Laudan, 1990b, 83–84, emphasis original).

13. To be honest, I am not sure how much am I at odds with Hacking or Pickering by this. The first invokes reality and truth every now and then. The latter heavily relies on the elevator attribute: "material."

14. A disclaimer is in place here. It is by no ways a self-evident task to describe a realist stance in the context of measurement (or anything else). My fictive realist opponent is my construction, of course. On the other hand, many theorists and practitioners are self-confessed realists, who do not have the urge to clearly state what they exactly mean on their realism. It is often insisted, however, that uncertainties are essential part of the result, not an assessment of it. I will touch on this point in a minute.

15. There is still a possibility that not every numerical assignment is measurement. In this case one should tell why uttering the sentence does not qualify as measurement. Anyhow, as no operationalist ever, to my knowledge, indulged in such speculations, we have no tracks to follow here.

16. We are not inclined to abandon it anyway, rather we tend to point the unreliability of our equipment.

17. As of December 2018 the power of the generators is estimated to be around 249 watts (NASA Jet Propulsion Laboratory 2018).

18. "In particular, the function of ad hoc corrections, rules of thumb and seemingly circular inferences prevalent in the production of UTC requires explanation. What role do these mechanisms play in stabilizing UTC, and is their use justified from an epistemic point of view? . . . I will consider two explanans that have been traditionally proposed for the stability of networks of physical measurement standards: (i) The empirical regularities exhibited by the behaviour of measurement standards (ii) The social coordination of policies for regulating and interpreting the behaviour of measurement standards" (Tal 2016, 8).

19. Such a theory is not an easy meat anyhow. Michael Dummett, an eminent one in this field, spent much of his life in his efforts for just clearing the way for such theories. The monumental and representative volume of Auxier and Hahn on Dummett's intellectual achievements nicely witness this fact (Auxier and Hahn 2007). On the thousand pages only a fragment of the texts could be said to be independent of the problem.

20. There is no definite limit for this series, as it would align with a curve. The place of the subsequent measuring points are random. But randomness is not the key point. Consider the simplest fractal. Take an equilateral triangle, and take middle third of each edge. Draw on these edge fractions equilateral triangles, now you got hexagram. Now iterate the operation on each newly made edge, and so on infinitely. Now, try to estimate the perimeter of the resulting object.

References

Albert, Réka, and Albert-László Barabási. 2002. "Statistical Mechanics of Complex Networks." *Reviews of Modern Physics* 74 (January).

Andréka, Hajnal, Judit X. Madarász, and István Neméti. 2002. *On the Logical Structure of Relativity Theories.* http: //www.renyi.hu/pub/algebraic-logic/olsort.html. Budapest: Alfréd Rényi Institute of Mathematics.

Andreski, Stanislav. 1972. *Social Sciences as Sorcery.* London: André Deutsch.

Auxier, Randall E., and Lewis Edwin Hahn, eds. 2007. *The Philosophy of Michael Dummett.* Peru, Illionis: Open Court Publishing Company.

Babbie, Earl Robert. 2007. *The Practice of Social Research.* Eleventh Edition. Thomson Wadsworth.

Bauer, Andrej. 2017. "Five Stages of Accepting Constructive Mathematics." *Bulletin of the American Mathematical Society* 54 (3): 481–498.

Berger, Peter L., and Thomas Luckmann. 1966. *The Social Construction of Reality: A Treatise in the Sociology of Knowledge.* Garden City, NY: Doubleday.

Bergmann, Gustav. 1956. "Sense and Nonsense in Operationism." In Frank 1956.

Berka, Karel. 1983. *Measurement: Its Concepts, Theories, and Problems.* D. Reidel Publishing Company.

Bickhard, Mark H. 2001. "The Tragedy of Operationalism." *Theory & Psychology* 11 (1): 35–44.

Bishop, Errett. 1967. *Foundations of Constructive Analysis.* New York: McGraw Hill Book Company.

Bloor, David. 1991 [1976]. *Knowledge and Social Imagery.* Chicago: University of Chicago Press.

Bridgman, Percy Williams. 1927. *The Logic of Modern Physics.* New York: Macmillan.

Brons, Lajos. 2005. *Rethinking the Culture: Economy Dialectic.* https://mpra.ub.uni-muenchen.de/162 5/1/MPRA_paper_1625.pdf. Munich Personal RePEc Archive.

Bulmer, M. 2001. "Social Measurement: What Stands in Its Way?" *Social Research,* no. 68: 455–480.

"SI brochure." 2006. Bureau International des Poids et Mesures (BIPM). www.bipm.org/utils/common/pdf/si_brochure_8_en.pdf.

Campbell, Norman Robert. 1920. *Physics: The Elements.* Cambridge University Press.

——— 1921. *What Is Science?* 36. Essex street W.C. London: Methuen & co. Ltd.

———. 1928. *An Account of the Principles of Measurement and Calculation.* London: Longmans, Green / Co.

Carnap, Rudolf. 1937. *The Logical Syntax of Language.* London: Kegan Paul.

———. 1945. "The Two Concepts of Probability." *Philosophy and Phenomenological Research* 5, no. 4 (June): 513–532.

———. 1950. *Logical Foundations of Probability.* Chicago: University of Chicago Press.

———. 1966. *An Introduction to the Philosophy of Science.* Dover edition, 1995. Edited by Martin Gardner. New York: Dover Publications, Inc.

Castellano, Daniel J. 2007. "Intensive and Extensive Magnitudes." http://www.arcaneknowledge.org/science/magnitude/magnitude.html.

Chang, Hasok. 2004. *Inventing Temperature: Measurement and Scientific Progress.* New York: Oxford University Press.

———. 2009. "Operationalism." In *The Stanford Encyclopedia of Philosophy*, Fall 2009, edited by Edward N. Zalta. http://plato.Stanford.edu/archives/fall2009/entries/operationalism/. Metaphysics Research Lab, Stanford University.

Chrisman, Nicholas R. 1998. "Rethinking Levels of Measurement for Cartography." *Cartography and Geographic Information Science* 4 (25): 231–242.

Churchman, C. West, and Philburn Ratoosh, eds. 1959. *Measurement, Definitions and Theories.* John Wiley / Sons, Inc.

CIPM. 2017. "International Committee for Weights and Measures - Proceedings of the 106th Meeting." 16–17 October. https://www.bipm.org/utils/en/pdf/CIPM/CIPM2017-EN.pdf.

Cliff, N. 1992. "Abstract measurement theory and the revolution that never happened." *Psychological Science*, no. 3:186–90.

Cohen, M. R., and Ernest Nagel. 1934. *An Introduction to Logic and Scientific Method.* London: Routledge / Kegan Paul.

Collier, David, and James E. Mahon. 1993. "Conceptual 'Stretching' Revisited: Adapting Categories in Comparative Analysis." *American Political Science Review* 87, no. 4 (December): 845–855.

Collins, H. M. 1981. "Son of Seven Sexes: The Social Destruction of a Physical Phenomenon." *Social Studies of Science* 11, no. 1 (February): 33–62.

Csatári, Ferenc. 2008. "A számfogalom matematikatörténeti fejlődéséről." *Szabad Változók,* no. 5.

———. 2012. "Some Remarks on the Physicalist Account of Mathematics." Doi: 10.4236/ojpp.2012.22025, *Open Journal of Philosophy*, no. 2:165–170.

———. 2016. "Measurement, Concepts, Sociology." DOI: 10.15476/ELTE.2016.162. PhD diss., Eótvós Loránd University.

Csatári, Ferenc, and Judit Juhász. 2009. "A PROMINSTAT migrációs metaadatbázis." *Statisztikai szemle*, nos. 87/7–8: 853–856.

Davis, Philip J., and Reuben Hersh. 1981. *The Mathematical Experience.* (References to the First Mariner Books edition, 1998.) Boston: Birkhäuser.

Dedekind, Richard. 1901. *Essays On the Theory of Numbers.* English by Wooster Woodruff Beman. Chicago: The Open Court.

Duhem, Pierre. 1906. *La Théorie Physique. Son Objet, sa Structure.* Paris: Chevalier & Riviére.

Dummett, Michael. 1991. *The Logical Basis of Metaphysics.* Harvard University Press.

———. 1993. "Realism." In *The Seas of Language.* Clarendon Press.

———. 1995. "Bivalence and vagueness." *Theoria*, no. 61: 201–216.

———. 2000. "Is Time a Continuum of Instants?" *Philosophy*, no. 75: 497–515.

Ellis, Brian. 1966. *Basic Concepts of Measurement.* Cambridge: Cambridge University Press.

Falk, Armin, Anke Becker, Thomas Dohmen, Benjamin Enke, David Huffman, and Uwe Sunde. 2018. "Global Evidence on Economic Preferences." *The Quarterly Journal of Economics* 133, no. 4 (November): 1645–1692.

Feest, Uljana. 2005. "Operationism in Psychology : What the Debate Is About, What the Debate Should Be About." *Journal of the History of the Behavioral Sciences* 41 (2): 131–149.

Feyerabend, Paul. 2010 [1975]. *Against Method.* (First Edition: New Left Books, London, 1975.) London: Verso.

Fodor, Jerry. 1974. "Special Sciences, or Disunity of Science as a Working Hypothesis." *Synthese* 2 (28): 97–115.

Forrest, Peter. 1995. "Is space-time discrete or continuous? An empirical question." *Synthese* 103 (3): 327–354.

Frank, Philipp G., ed. 1956. *The Validation of Scientific Theories.* Boston: Beacon Press.

Frege, Gottlob. 1884. *Die Grundlagen der Arithmetik - Eine logisch mathematische Unter- suchung uber den Brgiff der Zahl.* (References to the English edition: The Foundations of Arithmetic, Transleted by J. L. Austin. 1953. Harper & Brothers, New York.) Breslau: Verlag von Wilhelm Koebner.

Gerring, John. 1999. "What Makes a Concept Good? A Critical Framework for Understanding Concept Formation in the Social Sciences." *Polity* 31, no. 3 (Spring): 357–393.

Gillies, Donald A. 1972. "Operationalism." *Synthese*, no. 25:1–24.

Godfrey-Smith, Peter. 2014. *Philosophy of Biology.* Princeton: Princeton University Press.

Goodman, Nelson. 1955. *Fact, Fiction, and Forecast.* Fourth Edition, 1983. Harvard University Press.

Gould, Stephen Jay. 1996. *The Mismeasure of Man.* Revised Edition. W. W. Norton & Company.

Grace, Randolph C. 2001. "On the Failure of Operationism." *Theory & Psychology* 11 (1): 5–33.

Grégis, Fabien. 2019. "Assessing Accuracy in Measurement: The Dilemma of Safety versus Precision in the Adjustment of the Fundamental Physical constants." *Studies in History and Philosophy of Science Part A* 74:42–55.

Hacking, Ian. 1999. *The Social Construction of What?* Cambridge, MA: Harvard University Press.

Hannan, Michael T., László Polós, and Glenn R. Carroll. 2007. *Logics of Organization Theory: Audiences, Codes, and Ecologies.* Princeton University Press.

Hanson, Norwood Russell. 1958. *Patterns of Discovery.* Cambridge: Cambridge University Press.

Helmholtz, Hermann von. 1887. "Zählen und Messen, erkenntnis-theoretisch betrachet." *Philosophische Aufsatze Eduard Zeller gewidmet, Leipzig.*

Hempel, Carl Gustav. 1949 [1935]. "The Logical Analysis of Psychology." In *Readings in Philosophical Analysis,* edited by Herbert Feigl and Wilfrid Sellars, pp. 373–384. (Page references to a new edition: A Historical Introduction to the Philosophy of Mind: Readings with Commentary, Ed.: Peter A. Morton, 1997, Broadview Press Ltd., pp. 164–173. Original French edition: 1935.) New York: Appleton-Century-Crofts.

———. 1956. "A Logical Appraisal of Operationism." In Frank 1956.

———. 1966. *Philosophy of Natural Science.* Englewood Cliffs, NJ: Prentice-Hall.

Heyting, Arend. 1956. *Intuitionism: an Introduction.* Third Revised Edition 1971. Amsterdam: North-Holland Publishing Company.

Hölder, Otto. 1901. "Die Axiome der Quantität und die Lehre vom Mass." *Berichte über die Verhandlungen der Königlich Sächsischen Gesellschaft der Wissenschaften zu Leipzig, Mathematisch-Physische Klasse*, no. 53:1–46.

Hosiasson-Lindenbaum, Janina. 1940. "On Confirmation." *The Journal of Symbolic Logic* 5 (4): 133–148.

Huff, Darrell. 1954. *How to Lie with Statistics.* Penguin Books 1991. Victor Gollancz

"International Temperature Scale of 1990." 1990. http://www.its-90.com/.

Jeffreys, Harold. 1939. *Theory of Probability.* Oxford Press.

Kahneman, Daniel, and Amos Tversky. 1979. "Prospect Theory: An Analysis of Decision Under Risk." *Econometrica* 47 (263–291).

Kapsner, Andreas. 2014. *Logics and Falsifications.* Switzerland: Springer International Publishing.

Keynes, John Maynard. 1921. *A Treatise of Probability.* London: Macmillan.

Kolmogorov, A. N. 1932. "Zur Deutung der intuitionischen Logik." *Mathematische Zeitschrift,* no. 35.

Kootstra, Gerrit Jan. 2004. "Exploratory Factor Analysis: Theory and Application." http://www.let.rug.nl/-nerbonne/teach/rema-stats-meth-seminar/Factor-Analysis-Kootstra-04.PDF, Manuscript, Rijksuniversiteit Groningen.

Krantz, David H., Duncan R. Luce, Patrick Suppes, and Amos Tversky. 1971. *Foundations of Measurement.* Vol. Vol. I: Additive and Polynomial Representations. New York: Academic Press.

Kuhn, Thomas S. 1961. "The Function of Measurement in Modern Physical Science." *Isis* 52 (168): 161–193.

———. 1962. *The Structure of Scientific Revolutions.* Chicago, IL: University of Chicago Press.

Kyburg, Henry Ely. 1984. *Theory and Measurement.* Cambridge University Press.

———. 1991. "Measuring Errors of Measurement." In *Philosophical and Foundational Issues in Measurement Theory,* edited by C. Wade Savage and Philip Ehrlich. Psychology Press.

Lakatos, Imre. 1970. "Falsification and the Methodology of Scientific Research Programmes." In *Criticism and the Growth of Knowledge*, edited by Imre Lakatos and Alan Musgrave. Cambridge: Cambridge University Press.

———. 1976. *Proofs and Refutations: The Logic of Mathematical Discovery.* Cambridge, England: Cambridge University Press.

Latour, Bruno. 1987. *Science in Action: How to Follow Scientists and Engineers through Society.* Cambridge, MA: Harvard University Press.

Laudan, Larry. 1990a. "Demystifying Underdetermination." *Minnesota Studies in Philosophy of Science* 14:267–297.

———. 1990b. *Science and Relativism.* Chicago and London: University of Chicago Press.

Lord, Frederic M. 1953. "On the Statistical Treatment of Football Numbers." *American Psychologist,* no. 8: 750–751.

Luce, Duncan R., David H. Krantz, Patrick Suppes, and Amos Tversky. 1990. *Foundations of Measurement.* Vol. Vol. III: Representation, Axiomatization, and Invariance. New York: Academic Press.

Luce, Duncan R., and Louis Narens. 1981. "Axiomatic Measurement Theory." American Mathematical Society, *SIAM-AMS Proceedings* 13:213–235.

Mantzavinos, C., ed. 2009. *Philosophy of the Social Sciences, Philosophical Theory and Scientific Practice.* Cambridge University Press.

Margenau, Henry. 1956. "Interpretations and Misinterpretations of Operationalism." In Frank 1956.

———. 1959. "Philosophical Problems Concerning the Meaning of Measurement in Physics." In Churchman and Ratoosh 1959, 163–176.

Mari, Luca. 2003. "Epistemology of Measurement." *Measurement* 34 (1):17–30.

Maxwell, James Clerk. 1890. "General Considerations Concerning Scientific Apparatus." In: W. D. Niven (Ed.) The scientific papers of James Clerk Maxwell, vol. 2, pp. 505–522. New York: Dover, 1965.

Merton, Robert K. 1949. *Social Theory and Social Structure.* New York: Free Press.

Michell, Joel. 1990. *An Introduction to the Logic of Psychological Measurement.* Hillsdale, NJ: Lawrence Erlbaum Associates.

———. 1999. *Measurement in Psychology: A Critical History of a Methodological Concept.* Cambridge University Press.

———. 2007. "Measurement." In *Handbook of the Philosophy of Science. Philosophy of Anthropology and Sociology,* edited by Dov M. Gabbay, Paul Thagard, John Woods, Stephen P. Turner, and Mark W. Risjord, 71–119. Elsevier B.V.

Mises, Richard von. 1939. *Probability, Statistics and Truth.* New York: Macmillan.

Mosteller, Frederick, and John W. Tukey. 1977. *Data Analysis and Regression: a Second Course in Statistics.* Addison-Wesley.

Nagel, Ernest. 1931. "Measurement." *Erkenntnis,* no. 2: 313–335.

Narens, Louis. 1981b. "On the Scales of Measurement." *Journal of Mathematical Psychology,* no. 24: 249–275.

———. 2007. *Introduction to the Theories of Measurement and Meaningfulness and the Use of Symmetry in Science*. New Jersey: Lawrence Erlbaum Associates, Inc.

Narens, Louis, and Duncan R. Luce. 1993. "Further Comments On the 'Nonrevolution' of Axiomatic Measurement Theory." *American Psychological Society* 4, no. 2 (March).

NASA Jet Propulsion Laboratory. 2018. "Voyager." California Institute of Technology. December. https://voyager.jpl.nasa.gov/.

Neurath, Otto. 1932. "Protokollsätze." *Erkenntnis*, no. 3: 204–214.

Pap, Arthur. 1959. "Are Physical Magnitudes Operationally Definable." In Churchman and Ratoosh 1959, 177–191.

Papineau, David. 2009. "Physicalism and the Human Sciences." In Mantzavinos 2009, 103–123.

Pfanzagl, Johann. 1971. *Theory of Measurement.* 2nd Revised Edition. Berlin Heidelberg: Springer-Verlag GmbH.

Pickering, Andrew. 1984. *Constructing Quarks: A Sociological History of Particle Physics.* Chicago: University of Chicago Press.

———. 1995. *The Mangle of Practice: Time, Agency, and Science*. Chicago: University of Chicago Press.

Pluckrose, Helen, James A. Lindsay, and Peter Boghossian. 2018. "Academic Grievance Studies and the Corruption of Scholarship." Aero. October. https://areomagazine.com/2018/10/02/academic-grievance-studies-and-the-corruption-of-scholarship/.

Poincaré, Henri. 1902. *La Science et L'Hypothése.* Paris: Flammarion.

———. 1905. *La Valeur de la Science.* (Page references to the English edition: The Value of Science, Translated by George Bruce Halsted, 1907, The Science Press, New York.) Paris: Flammarion.

Popper, Karl. 1935. *Logik der Forschung.* (New English edition: The Logic of Scientific Discovery; Routledge Classics 2002; Taylor & Francis e-Library, 2005.) Verlag von Julius Springer, Vienna, Austria.

"PROMINSTAT Project." 2010. http://www.prominstat.eu/drupal/?q=node/64.

Quine, Willard V. 1951. "Two Dogmas of Empiricism." *The Philosophical Review* 60 (1):20– 43.

———. 1975. "On Empirically Equivalent Systems of the World." *Erkenntnis* 9 (3):313–328.

Reichenbach, Hans. 1949 [1935]. *Wahrscheinlichkeitslehre: eine Untersuchung über die logis-chen und mathematischen Grundlagen der Wahrscheinlichkeitsrechnung.* English translation: The Theory of Probability. Berkeley, California: University of California Press.

Reichheld, Frederick F. 2003. "One Number You Need to Grow." *Harvard Business Review* (December).

Resplandy, L., R. F. Keeling, Y. Eddebbar, M. K. Brooks, R. Wang, L. Bopp, M. C. Long, J. P. Dunne, W. Koeve, and A. Oschlies. 2018. "Quantification of Ocean Heat Uptake from Changes in Atmospheric O2 and CO2 Composition." *Nature* 563 (October): 105–108.

Roberts, Fred S. 1985. *Measurement Theory with Applications to Decisionmaking, Utility, and the Social Sciences.* Cambridge University Press.

Russell, Bertrand. 1903. *Principles of Mathematics.* (New edition: Routledge Classics, 2010; Taylor & Francis e-Library, 2009.) Cambridge: Cambridge University Press.

Sartori, Giovanni. 1970. "Concept Misformation in Comparative Politics." *American Political Science Review* LXIV(4):1033–1053.

———. 1984a. "Foreword." In Sartori 1984c.

———. 1984b. "Guidelines for Concept Analysis." In Sartori 1984c.

———, ed. 1984c. *Social Science Concepts: A Systematic Analysis.* Beverly Hills: Sage.

Scientific American. 2004. "What is the Fastest Event (Shortest Time Duration) That Can Be Measured with Today's Technology, and How Is This Done?" http: // www. scientificamerican.com/article/what-is-the-fastest-event/, *Scientific American* (December).

Shulman, R. G., and I. Shapiro. 2009. "Comment: Reductionism in the Human Sciences: A Philosopher's Game." In Mantzavinos 2009, 124–129.

Sokal, Alan, and Jean Bricmont. 1997. *Impostures Intellectuelles.* (In English: Fashionable Nonsense, Picador, 1998.) Éditions Odile Jacob.

Steele, Stephen F., and Jammie Price. 2007. *Applied Sociology. Terms, Topics, Tools, and Tasks.* 2nd Revised Edition. Wadsworth Publishing Co., Inc.

Stevens, Stanley Smith. 1935. "The Operational Definition of Psychological Concepts." *Psychological Review* 42:517–552.

———. 1946. "On the theory of scales of measurement." *Science*, no. 103: 677–680.

———. 1959. "Measurement, psychophysics and utility." In Churchman and Ratoosh 1959, 18–63.

Stigler, Stephen M. 1980. "Stigler's Law of Eponymy." (Gieryn, F., ed.) *Transactions of the New York Academy of Sciences* 39:147–158.

Stinchcombe, Arthur. 2005. *The Logic of Social Research.* Chicago: The University of Chicago Press.

Suppes, Patrick. 1951. "A Set of Independent Axioms for Extensive Quantities." *Portugaliae Mathematica*, no. 10:163–172.

———. 2002. *Representation and Invariance of Scientific Structures.* CSLI Publications, Center for the Study of Language / Information.

Suppes, Patrick, David H. Krantz, Duncan R. Luce, and Amos Tversky. 1989. *Foundations of Measurement, Vol. II: Geometrical, Threshold, and Probabilistic Representations.* New York: Academic Press.

Szabó, László E. 2002. *A nyitott jövő problémája.* Budapest: TYPOTEX Kiadó.

———. 2003. "Formal Systems as Physical Objects: A Physicalist Account of Mathematical Truth." *International Studies in the Philosophy of Science* 17:117–125.

———. 2009. "Empirical Foundation of Space and Time." In *EPSA07: Launch of the European Philosophy of Science Association*, edited by M. Suárez, M. Dorato, and M. Rédei. Springer.

———. 2013. "Vázlatpontok a fizikai elméletek fizikalista értelmezéséhez." In *Nehogy érvgyűlölők legyünk – Tanulmánykötet Máté András 60. születésnapjára*, edited by Zsófia Zvolenszky et al., 122–129. Budapest: L'Harmattan.

Tal, Eran. 2011. "How Accurate Is the Standard Second?" *Philosophy of Science* 78 (5): 1082–1096.

———. 2016. “Making Time: A Study in the Epistemology of Measurement.” *British Journal for the Philosophy of Science* 67:297–335.

Taylor, John R. 1997. *An Introduction to Error Analysis.* Second edition. Sausalito, California: University Science Books.

Teller, Paul. 2013. Measurement Accuracy Realism. Paper presented at Foundations of Physics 2013: The 17th UK and European Meeting on the Foundations of Physics.

“Trouble at the Lab.” 2013. October 19, 2013, *The Economist.*

“Out of the Box.” 2015. November 21, 2015, *The Economist*: pp. 55–56.

Thomson, James F. 1954. “Tasks and Super-Tasks.” *Analysis* 15, no. 1 (October):1–13.

Trendler, Günter. 2009. “Measurement Theory, Psychology and the Revolution That Cannot Happen.” *Theory & Psychology* 19 (5):579–599.

van Fraassen, Bas. 1980. *The Scientific Image.* Oxford: Oxford University Press.

Velleman, P. F., and L. Wilkinson. 1993. “Nominal, Ordinal, Interval, and Ratio Typologies are Misleading.” *The American Statistician* 47 (1).

Williams, Matt. 2016. “Planck Time.” http://www.universetoday.com/79418/planck-time/, Universe Today (April).

Zalta, Edward N. 2017. “Frege’s Theorem and Foundations for Arithmetic.” In *The Stanford Encyclopedia of Philosophy,* Summer 2017, edited by Edward N. Zalta. https://plato.stanford.edu/archives/sum2017/entries/frege-theorem/. Metaphysics Research Lab, Stanford University.

Index

About the Author

Ferenc Csatári, PhD, is an independent scholar. He earned his degree from Eötvös University, Budapest. His papers are concerned with topics in philosophy of mathematics, science, language, and social science methodology.

www.ingramcontent.com/pod-product-compliance
Ingram Content Group UK Ltd.
Pitfield, Milton Keynes, MK11 3LW, UK
UKHW041626031026
15587UKWH00001B/101

9 781498 582995